FOURIER ANALYSIS ON LOCAL FIELDS

BY

M. H. TAIBLESON

PRINCETON UNIVERSITY PRESS

AND

UNIVERSITY OF TOKYO PRESS

PRINCETON, NEW JERSEY

1975

Published by Princeton University Press, Princeton and London

L.C. Card: 74-32047

I.S.B.N.: 0-691-08165-4

Library of Congress Cataloging in Publication Data
will be found on the last printed page of this book

Published in Japan exclusively
by University of Tokyo Press
in other parts of the world by
Princeton University Press

Printed in the United States of America

For Charlotte

FOURIER ANALYSIS ON LOCAL FIELDS

By

M. H. Taibleson

Preface

These are the lecture notes of a course given at Washington University, Saint Louis during the Fall and Spring semesters 1972-73. With the exception of the results on Fourier series in II §6, all the results described have appeared in a series of papers, some of which were authored singly, but mostly they are the result of happy collaborations which started with Paul Sally. Then I went on to work with Richard Hunt, Keith Phillips and most recently with Jia-Arng Chao. Detailed references are given at the end of each chapter.

My particular thanks go to Richard Rochberg who audited the course, read the manuscript and has offered many valuable suggestions and to Paul Sally who bears special responsibility for getting me into all of this. Thanks also to my other collaborators: Hunt, Phillips and Chao for past support as well as help in preparing the manuscript, to Wei-Nung Liu who took notes of the lectures and especially to Mrs. Virginia Hundley for her superb typing of the manuscript.

Introduction

With the appearance of the paper of Gelfand and Graev [1] on representations of SL(2,K) and special functions on local fields, and the observation that most of the techniques used in that paper were, in form, rather straightforward extensions of well known methods in euclidean theory, Paul Sally and I set out on a program of developing and extending a few basic results that were needed for an elementary treatment of representation theory (Sally and Taibleson [1]). A review of related work in representation theory and algebraic number theory (See Lang's treatment of Tate's results, Lang [1], and other references in Sally and Taibleson [1]) indicated that such a development would be useful, in that a systematic treatment in a classical mold would make the results most understandable by virtue of the analogy with known material.

The program and its continuation, that was carried out and is described in these lectures, is a development of the basic facts about harmonic analysis on local fields and the n-dimensional vector spaces over these fields with an emphasis, almost exclusive, on the analogy between the local field and euclidean cases, with respect to the form of statements, the manner of proof and the variety of applications.

The force of the analogy rests in the relationship of the field structures which underlie the respective cases. A complete classification of locally compact, non-discrete fields gives us two examples of connected fields (the real and complex numbers) and the rest which are local fields (the p-adic numbers, finite algebraic extensions of the p-adic numbers and formal series fields with coefficients from a finite field). We are, in a certain sense, studying local fields as examples that round out what we know about the reals and complexes as locally compact fields; which is to say, as analytic objects in which the algebraic operations of addition and multiplication interact.

Our point of view has its virtues, but it is somewhat narrow and not totally satisfactory. To illustrate, an interesting subject of study is Fourier analysis on $\mathfrak{D}$, the ring of integers in the local field K. $\mathfrak{D}$ is the maximal compact subring in K. From our point of view $\mathfrak{D}$ is a subring of K and this view is useful, particularly since harmonic analysis on $\mathfrak{D}$ is the analogue of the study of Fourier series on the torus, $\mathbb{T}$. Note that $\mathbb{T} \cong \mathbb{R}/\mathbb{Z}$ where $\mathbb{Z}$ is the ring of rational integers and $\hat{\mathbb{T}} = \mathbb{Z}$; and in the case of a local field, the role of the integers is played by $\mathfrak{D}$, a compact abelian group, and $\hat{\mathfrak{D}} \cong K/\mathfrak{D}$, a countable discrete group.

One could just as well have studied $\mathfrak{D}$ as a Vilenkin group (see Vilenkin [1] and for examples of more recent work, Onneweer [1] and

Waterman [1]); that is, as a compact abelian, separable group with a periodic dual, and concentrate on the consequences of these facts; or we could note that $\mathfrak{D}$ is a probability space and there is a natural martingale operator on $\mathfrak{D}$ which can be used to study the behaviour of functions on $\mathfrak{D}$. In fact, there is a classical object, 2^{ω} (the dyadic group or Walsh-Paley group) which has been the subject of intensive probabilistic study, and which turns out to be the additive group of the ring of integers in the 2-series field. (See Burkholder [1], [2] and Gundy [1], [2], for some basic results on martingales and applications to the dyadic group.)

We have made a choice to take the narrower point of view and stress the analogy between harmonic analysis on local fields and euclidean analysis; developments with emphasis on the elementary topological-algebraic structures or probabilistic structures will be left to other authors and/or other times.

All formal references are contained in the ultimate section of each chapter. It would be helpful if these notes were read along with the main text.

This presentation has been motivated by a desire to lay out the basic facts of Fourier analysis on local fields in an accessible form, and in the same spirit as in the treatise on trigonometric series by

Zygmund [1] or more recently by Edwards [1]; and by Stein and Weiss [1] and Stein [1] for Fourier analysis on euclidean spaces.

A suggested prerequisite for reading these lecture notes is that usual object, a good graduate course in Lebesgue integration and functional analysis. For maximum benefit the serious student is urged to read these notes with Stein and Weiss [1] and Stein [1] immediately at hand for comparison with the euclidean case.

Table of Contents

Chapter I: Introduction to local fields

This chapter contains a description of several examples of local fields and a review of basic facts about the classification and structure of local fields.

1. Rademacher functions and the Walsh-Paley group

The Rademacher functions are defined on $[0,1]$ by the rule, $\varphi_k(x) = \operatorname{sgn}(\sin 2^{k+1}\pi x)$, $k=1,2,3,\cdots$. It is easy to see that the sequence $\{\varphi_k\}_{k=1}^{\infty}$ is orthonormal on $[0,1]$ with respect to ordinary Borel-Lebesgue measure. In the language of probability theory we would say that the sequence of Rademacher functions is a sequence of uncorrelated random variables, each with mean zero and variance 1. While it is not crucial for the development that follows, it is of interest to note that these functions are also equi-distributed and independent.

$\{\varphi_k\}_{k=1}^{\infty}$ is not complete. It is extended to a complete system as follows: Let $\psi_0(x) \equiv 1$, and if n is a positive integer we write n uniquely as, $n = a_0 + a_1 2 + \cdots + a_\ell 2^\ell$, where $a_\ell = 1$ and $a_k = 0$ or 1, $k = 0,1,\cdots,\ell-1$. Then we let $\psi_n = \prod_{k=0}^{\ell}[\varphi_k]^{a_k}$. Thus, $\varphi_k = \psi_{2^k}$, $k = 1,2,\cdots$. The system $\{\psi_n\}_{n=0}^{\infty}$ is a complete orthonormal sequence on $[0,1]$. It is known as the Walsh-Paley system.

For $f \in L^1[0,1]$ we define the Fourier coefficients and partial sums of the Fourier series of f with respect to the Walsh-Paley system in the usual way. That is, we let $c_k = \int_0^1 f(x)\psi_k(x)dx$, $k = 0,1,\cdots$, and $S_n(x;f) = \sum_{k=0}^{n-1} c_k\psi_k(x)$. These partial sums have a most curious property. Let $x \in [0,1]$, $I_{k,x}$ be the dyadic interval of length 2^{-k} that contains x and assume for the sake of simplicity that x is not a dyadic rational. Then $S_{2^k}(x;f) = 2^k \int_{I_{k,x}} f(z)dz = \frac{1}{|I_{k,x}|} \int_{I_{k,x}} f(z)dz$

This phenomenon is explained by the following: If $x \in [0,1]$ we may write $x = \sum_{k=0}^{\infty} a_k 2^{-k}$, $a_k = 0$ or 1. There is also a space of sequences $\{a_0,a_1,\ldots\}$ of zeros and ones. We identify x with the sequence $\{a_0,a_1,\ldots\}$. If x is a dyadic rational, $0 < x < 1$, then this representation is not unique, but it only fails to be 1:1 on this countable set. In this sequence space an addition, usually denoted $x \dotplus y$, is defined as addition of sequences coordinatewise, mod 2. When this collection of sequences is given the obvious metric topology, it is easily seen that it is a compact abelian group. This group is called the dyadic group, 2^ω, or the Walsh-Paley group. A remarkable fact is that the Borel ring and Haar measure on 2^ω agree with the usual Borel-Lebesgue structure of $[0,1]$.

The sequence of Walsh-Paley functions is a complete set of characters on 2^ω.

2. The 2-adic numbers, and 2-series numbers

The 2-adic norm, $|\cdot|_2$, is defined on the (rational) integers as follows: $|0|_2 = 0$. If $n \neq 0$ we write $n = 2^k a$ where $(a,2) = 1$ and let $|n|_2 = 2^{-k}$. It is easy to see that $x \to |x|_2$ is a norm on the integers. Moreover, $|n+m|_2 \leq \max[|n|_2, |m|_2]$ for n and m integers. If we keep the usual arithmetic for the integers and define a metric by $d(n,m) = |n-m|_2$ we see that, with this metric, the integers form a metric space. Its completion is called the ring of 2-adic integers. It is easily identified with the collection of formal power series $\{x = \sum_{k=0}^{\infty} a_k 2^k\}$ where each $a_k = 0$ or 1, and where addition and multiplication is done formally, carrying from left to right. In this collection we see that $|x|_2 = 2^{-\ell}$ iff $x = 2^{\ell} + \sum_{k=\ell+1}^{\infty} a_k 2^k$. We also see that the non-negative (rational) integers are exactly that collection of 2-adic integers with finite expansions. It is a useful exercise to verify that $-1 = 1 + 2 + 2^2 + \cdots = 1/(1-2)$. (Note that $|2|_2 = 1/2$.)

It is easily verified that the ring of 2-adic integers, under addition, is a compact abelian group. Its field of quotients is called the 2-adic numbers. The 2-adic numbers can also be obtained by extending the 2-adic norm to the rational numbers (write $n/m = 2^k \cdot a/b$

with a and b relatively prime to 2 and set $|n/m|_2 = 2^{-k}$) and then completing the rationals as a metric space with the induced metric. In either case we obtain a locally compact topological field of characteristic zero, that is totally disconnected. Its elements are identified as formal Laurent series, $\{x = \sum_{k=\ell}^{\infty} a_k 2^k\}$, where each $a_k = 0$ or 1 and we "carry" in the arithmetic. Note also that if $a_\ell = 1$ in the representation above, then $|x| = 2^{-\ell}$.

If we consider the same set of formal symbols, but now do the addition and multiplication with the coefficients $\{a_k\}$ viewed as elements of GF(2), and use the same topology and norm as we used for the 2-adic numbers, we also obtain a locally compact topological field that is totally disconnected. However, this field is of characteristic 2. It is called the 2-series field. The set of power series elements in the 2-series field (namely the collection of elements $x = \sum_{k=0}^{\infty} a_k p^k$) is seen to be a compact ring. It is not difficult to verify that the power series ring of the 2-series field coincides, under addition, with the Walsh-Paley group, 2^ω. The group 2^ω is thus endowed, in a natural way, with a multiplication.

3. p-adic and p-series numbers

In a manner that is directly analogous to that for 2-adic and 2-series numbers we may construct p-adic and p-series fields where

p is any rational prime. These fields will be locally compact fields that are totally disconnected and are either of characteristic zero (p-adic) or of characteristic p (p-series).

4. Classification of local fields

Let K be a field and a topological space. Then K is a locally compact field if K^+ and K^* are locally compact abelian groups (where K^+ and K^* are the additive and multiplicative groups of K).

If K is any field and is endowed with the discrete topology then K is a locally compact field. Therefore we exclude the discrete topological fields from consideration. (The only "naturally" discrete fields are the finite fields.)

We consider fields K that are locally compact, non-discrete and (redundantly but emphatically) topologically complete.

If K is connected then K is either $\mathbb{R}$ or $\mathbb{C}$.

If K is not connected then K is totally disconnected.

If K is of finite characteristic, then K is a field of formal power series over a finite field $GF(p^c)$. If $c = 1$ it is a p-series field. If $c \neq 1$ then K is an algebraic extension of degree c of a p-series field.

If K is of characteristic zero then K is either a p-adic number field or a finite algebraic extension of such a field.

In the sequel, a local field K is always a locally compact, non-discrete, totally disconnected field.

5. Properties of local fields

Let K be a fixed local field. Since K^+ is a locally compact abelian group we may choose a Haar measure dx for K^+. If $\alpha \neq 0$ $(\alpha \in K)$ then $d(\alpha x)$ is also a Haar measure. Let $d(\alpha x) = |\alpha| dx$ and call $|\alpha|$ the absolute value or valuation of α. Let $|0| = 0$.

Comment. The mapping $\alpha \to |\alpha|$ is simply the modular function of the endomorphism $x \to \alpha x$.

The mapping $x \to |x|$ has the following properties: $|x| = 0$ iff $x = 0$; $|xy| = |x||y|$; and $|x+y| \leq \max[|x|,|y|]$. The last of these properties is called the ultrametric inequality and a norm or valuation which satisfies it is said to be ultrametric or a non-archimedian norm.

Note. If $|x| \neq |y|$ then $|x+y| = \max[|x|,|y|]$.

Proof. Suppose $|x| > |y|$ and $|x+y| < |x|$. Then $|x| = |(x+y) - y| \leq \max[|x+y|,|y|] < |x|$, a contradiction.

Fact. A Haar measure on K^* is $\frac{dx}{|x|}$.

Notation. $\mathfrak{O} = \{x \in K : |x| \leq 1\}$. $\mathfrak{O}$ is called the ring of integers in K.

The ring of integers in a local field K is the unique maximal compact subring of K.

Example. If K is the 2-adic number field then $\mathfrak{O}$ is the ring of 2-adic integers. If K is the 2-series field then the additive group of $\mathfrak{O}$ is the Walsh-Paley group.

Notation. $\mathfrak{P} = \{x \in K : |x| < 1\}$. $\mathfrak{P}$ is called the prime ideal in K.

The prime ideal in K is the unique maximal ideal in $\mathfrak{O}$. It is principal and prime. The fact that K is totally disconnected implies that the valuation is discretely valued. That is, the set of absolute values is of the form $\{s^k\}_{k=-\infty}^{+\infty} \cup \{0\}$ for some $s > 0$. Thus, there is an element of $\mathfrak{P}$ of maximum absolute value.

Notation. Let $\mathfrak{p}$ be a fixed element of maximum absolute value. $\mathfrak{p}$ is called a prime element of K. As an ideal in $\mathfrak{O}$, $\mathfrak{P} = (\mathfrak{p}) = \mathfrak{p}\mathfrak{O}$.

Example. If K is the 2-adic numbers as defined in §2 then $x \in \mathfrak{O}$ iff $x = \sum_{k=0}^{\infty} a_k 2^k$, $a_k = 0$ or 1; $x \in \mathfrak{P}$ iff $x = \sum_{k=1}^{\infty} a_k 2^k$, $a_k = 0$ or 1; and $\mathfrak{p} = 2$ ($\mathfrak{p}^0 = 1$).

Fact. $\mathfrak{O}$ is compact and open. Hence $\mathfrak{P}$ is compact and open.

It follows from these facts that $\mathfrak{O}/\mathfrak{P}$ is isomorphic to a finite field $GF(q)$, where $q = p^c$ for some prime p and positive integer c. To see this note that $\mathfrak{P}$ compact implies that $\mathfrak{O}/\mathfrak{P}$ is compact, that $\mathfrak{P}$ open implies that $\mathfrak{O}/\mathfrak{P}$ is discrete and since $\mathfrak{P}$ is maximal $\mathfrak{O}/\mathfrak{P}$ is a field and hence a finite field. In the sequel we assume that the prime power, $q = p^c$ is fixed.

<u>Notation.</u> For E a measurable subset of K let $|E| = \int_K \xi_E(x)dx$ where ξ_E is the characteristic function of E and dx is a Haar measure normalized so $|\mathfrak{O}| = 1$.

<u>Facts</u>. $|\mathfrak{P}| = q^{-1}$. $|\mathfrak{p}| = q^{-1}$.

<u>Proof</u>. Decompose $\mathfrak{O}$ into q cosets of $\mathfrak{P}$. Thus $1 = |\mathfrak{O}| = q|\mathfrak{P}|$, and so $|\mathfrak{P}| = q^{-1}$. But $\mathfrak{P} = \mathfrak{p}\mathfrak{O}$. Thus $q^{-1} = |\mathfrak{P}| = |\mathfrak{p}||\mathfrak{O}| = |\mathfrak{p}|$.

It follows that if $x \in K$ and $x \neq 0$ then $|x| = q^k$ for some $k \in \mathbb{Z}$.

<u>Notation</u>. $\mathfrak{O}^* = \mathfrak{O} \sim \mathfrak{P} = \{x \in K: |x| = 1\}$. $\mathfrak{O}^*$ is the <u>group of units</u> in K^*. If $x \neq 0$ we may write $x = \mathfrak{p}^k x'$ with $x' \in \mathfrak{O}^*$.

<u>Example</u>. For K the 2-adic integers, $\mathfrak{O}^* = \{\sum_{k=0}^{\infty} a_k 2^k : a_0 = 1\}$.

<u>Notation</u>. $\mathfrak{P}^k = \mathfrak{p}^k\mathfrak{O} = \{x \in K: |x| \leq q^{-k}\}$, $k \in \mathbb{Z}$. $\mathfrak{P}^k$ is called a <u>fractional ideal</u>.

Each $\mathfrak{P}^k$ is compact and open and is a subgroup of K^+. If $k \geq 0$ then $\mathfrak{P}^k$ is a subring of K. The collection $\{\mathfrak{P}^k\}_{k\in\mathbb{Z}}$ is a fundamental system of neighborhoods of 0.

<u>Notation</u>. Let $A_0 = \mathfrak{O}^*$, $A_k = 1 + \mathfrak{P}^k \subset \mathfrak{O}^*$, $k = 1,2,\cdots$.

The collection $\{A_k\}_{k=0}^{\infty}$ is a fundamental system of neighborhoods of 1 in $\mathfrak{O}^*$.

A few miscellaneous facts about K.

A. *If* $a \in K$ *then* $\lim_{n\to\infty} a^n = 0$ *iff* $|a| < 1$.

B. *If* L *is a discrete subfield of* K *then* L *is finite*.

From A we see that if $x \in L$ then $|x| = 1$ or $|x| = 0$. Thus, $L \subset \mathfrak{O}^*$ and being a discrete subset of a compact set, it is finite.

C. *Let* V *be a topological vector space over* K. *Let* V′ *be a finite dimensional subspace of* V *with basis* $\{v_1, \ldots, v_n\}$. *Then the map* $(x_1, \ldots, x_n) \to \sum x_k v_k$ *of* K^n *to* V′ *is an isomorphism of* K^n *and* V′ *as topological vector spaces, where* K^n *is given the topological structure induced by the norm* $|(x_1, \ldots, x_n)| = \sup_k |x_k|$.

D. *Every finite dimensional vector space over* K *can be given only one structure as a topological vector space*.

E. *If* V *is a locally compact topological vector space over* K *then* V *has finite dimension* d, *and* $\mathrm{mod}_V(x) = |x|^d$ *for all* $x \in K$.

F. *Let* A *be an endomorphism of a vector space*, V, *of finite dimension over* K. *Then* $\mathrm{mod}_V(A) = |\det A|$.

Theorem (A summary).

Let K *be a local field*, $\mathfrak{O} = \{x \in K: |x| \le 1\}$, $\mathfrak{O}^* = \{x \in K: |x| = 1\}$, *and* $\mathfrak{P} = \{x \in K: |x| < 1\}$. *Then* K *is ultrametric*, $\mathfrak{O}$ *is the unique maximal compact subring in* K, $\mathfrak{O}^*$ *is the group of units in* K^*, *and* $\mathfrak{P}$ *is the unique maximal ideal in* $\mathfrak{O}$. *There is a* $\mathfrak{p} \in \mathfrak{P}$ *such that*

$\mathfrak{P} = \mathfrak{p}\mathfrak{O}$. *The residue space* $\mathfrak{k} = \mathfrak{O}/\mathfrak{P}$ *is a finite field of characteristic* p. *If* q *is the number of elements in* $\mathfrak{k}$ *then the image of* K^* *in* $(0,\infty)$ *under the valuation* $|\cdot|$ *is the subgroup of* $(0,\infty)$ *generated by* q. $|\mathfrak{p}| = q^{-1}$. *A Haar measure on* K^* *is given by* $\dfrac{dx}{|x|}$.

6. More facts about K

A. *Let* $\{a_k\}_{k=0}^{\infty}$ *be any sequence in* K *with limit* 0 *in* K. *Then,* $\sum_{k=0}^{\infty} a_k$ *converges commutatively.*

An easy consequence of the ultrametric inequality.

B. *Let* $\mathfrak{A} = \{a_i\}_{i=1}^{q}$ *be any fixed full set of coset representatives of* $\mathfrak{P}$ *in* $\mathfrak{O}$. *Then if* $x \in \mathfrak{P}^k$, $k \in \mathbb{Z}$, x *can be expressed uniquely as* $x = \sum_{\ell=k}^{\infty} c_\ell \mathfrak{p}^\ell$, $c_\ell \in \mathfrak{A}$.

Proof. We may assume $k = 0$. The c_ℓ are defined inductively by the relations $x \equiv \sum_{\ell=0}^{N} c_\ell \mathfrak{p}^\ell$, $\mathrm{mod}(\mathfrak{P}^{N+1})$.

C. *Let* A *be an automorphism of* K (*as a topological field*). *Then* A *maps* $\mathfrak{O}$ *onto* $\mathfrak{O}$, $\mathfrak{P}$ *onto* $\mathfrak{P}$ *and has module* 1 *as an automorphism of* K^+.

Proof. Clearly the image of $\mathfrak{O}$ is a maximal compact subring of K and so $A\mathfrak{O} = \mathfrak{O}$. Similarly $A\mathfrak{P} = \mathfrak{P}$. Now note that $\mathrm{mod}_K(A)|\mathfrak{O}| = |A\mathfrak{O}| = |\mathfrak{O}|$ so $\mathrm{mod}_K(A) = 1$.

D. Let A be as in C. Then for all $x \in K$, $|Ax| = |x|$.

Proof. From C it follows that A maps $\mathfrak{D}^* = \mathfrak{D} \sim \mathfrak{P}$ onto itself, so $(|x| = 1) \Rightarrow (|Ax| = 1)$. Since $A\mathfrak{p}$ generates $A\mathfrak{P} = \mathfrak{P}$ we must have $|A\mathfrak{p}| = q^{-1}$. If $x \neq 0$ write $x = \mathfrak{p}^k x'$, $x' \in \mathfrak{D}^*$. Then $|Ax| = |A\mathfrak{p}|^k \cdot |Ax'| = q^{-k} \cdot 1 = |x|$. If $x = 0$ then $Ax = 0$ and again $|Ax| = |x|$.

E. Let K be a local field. Then K^* contains a subgroup of order $q-1$ which is cyclic and unique. Let M^* be that subgroup. M^* is the group of roots of 1 that are of order prime to p, which is to say the roots of $x^{q-1} = 1$. $M = M^* \cup \{0\}$ is a full set coset representatives of $\mathfrak{P}$ in $\mathfrak{D}$. If K is of characteristic p then $M = GF(q)$.

7. The dual of K^*

Let ϵ be a generator of M^*. Then, since $\{\epsilon^\ell\}_{\ell=0}^{q-2} \cup \{0\}$ is a complete set of coset representatives of $\mathfrak{P}$ in $\mathfrak{D}$ we have that if $x \in K^*$ then x can be written uniquely: $x = \mathfrak{p}^k \epsilon^\ell a$, where $a \in A = 1 + \mathfrak{P}$, $k \in \mathbb{Z}$, $\ell = 0, 1, \ldots, q-2$.

k is determined by $|x| = q^{-k}$, ℓ by $\mathfrak{p}^{-k} x \equiv \epsilon^\ell \pmod{\mathfrak{P}}$, and it is obvious that $\mathfrak{p}^{-k}\epsilon^{-\ell} x = a \in A$.

It is now simple to see that $\mathfrak{D}^* = \mathbb{Z}_{q-1} \times A$ and that $K^* \cong \mathbb{Z} \times \mathfrak{D}^* \cong \mathbb{Z} \times \mathbb{Z}_{q-1} \times A$, where $\mathbb{Z} \cong \{\mathfrak{p}^k\}_{k \in \mathbb{Z}}$, $\mathbb{Z}_{q-1} \cong \{\epsilon^\ell\}_{\ell=0}^{q-2} = M^*$ and $A = 1 + \mathfrak{P}$.

Noting that $\hat{\mathbb{Z}} = \mathbb{T}$ (the circle group), $\hat{\mathbb{Z}}_{q-1} = \mathbb{Z}_{q-1}$ (the cyclic group of order q-1) and (since A is compact and separable) $\hat{A}$ is a countable discrete group we see that $\hat{K}^* = \mathbb{T} \times \mathbb{Z}_{q-1} \times \hat{A}$, so that $\hat{K}^*$ has the structure of a countable discrete collection of circles.

Let π be a (continuous) unitary multiplicative character on K^* (that is, $\pi \in \hat{K}^*$). [In the sequel we will assume that multiplicative characters are continuous, <u>but we do not assume in general that they are unitary</u>.] Then π can be decomposed uniquely $\pi(x) = \pi(x')|x|^{i\alpha}$, where $-\pi/\ln q < \alpha \leq \pi/\ln q$. (Recall that $|x| = q^k$ for some k.)

We denote the mapping $x \to \pi(x')$ by $\pi^*(x)$. π^* can be viewed as a character on $\mathfrak{O}^*$ or as a character on K^* that is homogeneous of degree zero in the sense that $\pi^*(\mathfrak{p}^k x) = \pi^*(x)$ for all $k \in \mathbb{Z}$. We can further decompose π^* by writing $x' = \epsilon^{\ell} a$, so $\pi^*(x') = (\pi^*(\epsilon))^{\ell} \pi^*(a)$, so that π^* is determined by the value of $\pi^*(\epsilon)$ and its behaviour on A.

<u>Proposition</u>. <u>If</u> $\pi \in \hat{K}^*$ <u>then</u> π <u>is equal to</u> 1 <u>on some</u> A_k, $k = 0,1,\cdots$.

<u>Proof</u>. Since π is continuous and $\{A_k\}_{k=0}^{\infty}$ is a fundamental system of neighborhoods of 1 in K^*, there is a $k \in \mathbb{Z}$ such that if $x \in A_k$ then $|\pi(x) - 1| < \sqrt{2}$. Since A_k is a multiplicative group, $x^{\ell} \in A_k$ for any positive integer ℓ, $\pi(x^{\ell}) = (\pi(x))^{\ell}$, so $|(\pi(x))^{\ell} - 1| < \sqrt{2}$ for all such ℓ. This is not possible unless $\pi(x) =$

__Notation__. If $\pi(x) = 1$ for all $x \in A_0 = \mathfrak{O}^*$ we say that π is __unramified__. Otherwise we say that π is __ramified__. If π is ramified then the smallest integer k such that π is constant on A_k is called the __degree of ramification__ of π and we say that π is __ramified of degree__ k.

8. The dual of K^+

__If__ K __is a local field there is a non-trivial, unitary, continuous character__, χ, __on__ K^+. [In the sequel we always assume that additive characters are continuous and unitary.] The existence of such a character follows from the Pontryagin Duality Theorem, though in particular cases it is easy to construct such characters directly.

__Examples__. 1) Let K be the 2-series field. Write $x \in K$ as $x = x_0 + \sum_{k=\ell}^{-1} a_k \mathfrak{p}^k$, $a_k = 0$ or 1, $x_0 \in \mathfrak{O}$. Define $\chi(\mathfrak{p}^k) = \begin{cases} -1, & \text{if } k = -1 \\ 1, & \text{if } k < -1 \end{cases}$, $\chi(x_0) = 1$. χ is a character on K^+ that is trivial on $\mathfrak{O}$, but is non-trivial on $\mathfrak{P}^{-1}$.

2) Let K be the 2-adic field and write $x \in K$ as

$$x = x_0 + \sum_{k=\ell}^{-1} a_k 2^k, \quad a_k = 0 \text{ or } 1, \quad x_0 \in \mathfrak{O}.$$

Define $\chi(2^k) = e^{2\pi i 2^k}$, $k \leq -1$ and $\chi(x_0) = 1$. χ is again a character on K^+ that is trivial on $\mathfrak{O}$, but is non-trivial on $\mathfrak{P}^{-1}$.

<u>Proposition</u>. <u>If</u> χ <u>is a character on</u> K^+ <u>then there is a</u> $k \in \mathbb{Z}$ <u>such that</u> χ <u>is trivial on</u> $\mathfrak{P}^k$.

<u>Proof</u>. We argue as in the Proposition in §7. Since χ is continuous there is a $k \in \mathbb{Z}$ such that $|\chi(x) - 1| < \sqrt{2}$ for all $x \in \mathfrak{P}^k$. For all positive integers ℓ, $\ell x \in \mathfrak{P}^k$ and $\chi(\ell x) = (\chi(x))^\ell$, so $|(\chi(x))^\ell - 1| < \sqrt{2}$ for $\ell = 1,2,\cdots$, which is impossible unless $\chi(x) = 1$.

<u>Remark</u>. If χ is trivial on $\mathfrak{P}^k$ then χ is constant on cosets of $\mathfrak{P}^k$ in K^+. For if $x \in y + \mathfrak{P}^k$, so $x = y + z$, $z \in \mathfrak{P}^k$, then $\chi(x) = \chi(y)\chi(z) = \chi(y)$.

<u>Proposition</u>. <u>Let</u> χ <u>be a non-trivial character on</u> K^+. <u>Then the correspondence</u> $\lambda \longleftrightarrow \chi_\lambda$ <u>where</u> $\chi_\lambda(x) = \chi(\lambda x)$ <u>establishes a topological isomorphism of</u> K^+ <u>with a one-dimensional sub-space of</u> $\hat{K}^+$.

<u>Proof</u>. Left as an exercise.

<u>Theorem</u>. K^+ <u>is self dual</u> $(\hat{K}^+ \cong K^+)$. <u>The correspondence is established by the mapping</u> $\lambda \longleftrightarrow \chi_\lambda$, <u>where</u> χ <u>is any non-trivial character on</u> K^+.

<u>Proof</u>. One first checks that $\hat{K}$ is a locally compact vector space over with vector addition defined by $(\chi_1,\chi_2) \to \chi_1\chi_2[(\chi_1\chi_2)(x) = \chi_1(x)\chi_2(x)]$ and scalar multiplication defined by $(\lambda, \chi) \to \chi_\lambda[\chi_\lambda(x) = \chi(\lambda x)]$.

By E of §5 $\hat{K}$ has finite dimension d. That is $\hat{K} \cong K^d$. But $K \cong (\hat{K})^\wedge \cong K^{d^2}$, which implies that $d = 1$ and that $\hat{K} \cong K$.

<u>Corollary</u>. <u>If</u> F <u>is a locally compact, non-discrete field then</u> F <u>is self dual</u>.

F totally disconnected is the Theorem immediately above. The only other cases are $F = \mathbb{R}$ or $\mathbb{C}$. Actually the result extends to division rings with only modest modification.

For $x, y \in K^n$, $x = (x_1, \ldots, x_n)$, $y = (y_1, \ldots, y_n)$ we define $x \cdot y = x_1 y_1 + \cdots + x_n y_n$.

<u>Definition</u>. For $x, y \in K^n$, $\chi_y(x) = \chi(y \cdot x)$, for $\chi \in \hat{K}^+$.

We see that χ_y is a character on K^n whenever $\chi \in \hat{K}^+$.

<u>Corollary</u>. <u>Let</u> K <u>be a local field</u>, K^n <u>an n-dimensional vector space over</u> K. <u>Then</u> $\hat{K}^n \cong K^n$ <u>where the correspondence is established by the relation</u> $y \longleftrightarrow \chi_y$.

In the sequel χ is a fixed character on K^+ that is trivial on $\mathfrak{O}$ but is non-trivial on $\mathfrak{P}^{-1}$. It follows that χ is constant on cosets of $\mathfrak{O}$ and that if $y \in \mathfrak{P}^k$ then χ_y is constant on cosets of $\mathfrak{P}^{-k}$. We find χ by starting with any non-trivial character and rescaling.

9. Notes for Chapter I

General references for this chapter are Chapters 5 and 6 of Bourbaki [1], the opening sections of Gelfand and Graev [1] and the introduction of Sally and Taibleson [1].

§1. The Rademacher functions are important objects of study in real and complex analysis. See, for instance, Zygmund [1, vol. I, Ch. 5 & 8 for examples and further references. Fundamental papers on the dyadic group and the Walsh-Paley functions are the papers of Paley [1], Walsh [and Fine [1].

It would seem that, initially, much of the motivation for the stud of the Walsh-Paley system was a desire to understand functions on [0,1] as they were approximated by functions that are constant on dyadic intervals. From that point of view the Walsh system had a strong competitor; namely, the Haar system, which is also a complete orthonormal system on [0,1]. See the papers of Paley and Walsh for detail

The "obvious metric" referred to towards the end of §1 is, of course, the product metric.

§4. The classification of local fields is carried out in detail in the first chapter of Weil's book [1], where he also gives a classification of skew p-fields. What we have termed local fields are called by Weil, p-fields.

§5. The reference to the "modular function" can use a modest amount of explanation. If G is a unimodular (locally compact) group with Haar measure μ, and A is an endomorphism of G, then either $E \rightarrow \mu(AE)$ is a Haar measure on G, or $\mu(AE) \equiv 0$ for measurable sets E in G. In either case, the relation $\mu(AE) = \mathrm{mod}_G(A)\mu(E)$ well defines $\mathrm{mod}_G(A)$, which is called the <u>module</u> or <u>modular function</u> of the endomorphism A. If K is a topological field (skew or commutative) then $\mathrm{mod}_K(\alpha) \equiv |\alpha|_K \equiv |\alpha|$, $\alpha \in K$ is the module of the endomorphism of K^+, $x \rightarrow \alpha x$. If V is a locally compact vector space over K and $x \in K$, $\mathrm{mod}_V(x)$ is the module of "scalar multiplication by x". If V is a finite dimensional vector space over K then A, an endomorphism of V, can be identified with a linear transformation of V and the determinant of that linear transformation is an invariant of A. In that sense, $\mathrm{mod}_V(A) = |\det A|$. The proof is standard.

It is not difficult to see that the modular map $x \rightarrow |x|$ is continuous and defines the topology of K. It is also easy to see that for the p-adic and p-series fields it agrees with the norms defined in §3 and §4.

§6. The locution "converges commutatively" is a substitute for "converges absolutely" which is used for complex or real series. It means, of course, that all rearrangements converge, and to the same limit. If you want to get fancy, "the Lebesgue integral of the K-value function on the non-negative integers (with mass 1 concentrated at each integer) exists" carries much the same content.

§7, §8. The reference to $\hat{K}^*$ or $\hat{K}^+$, as the dual of K^* or K^+, is to the Pontryagin dual of the given group, as a locally compact abelian group (see Hewitt and Ross [1] for details). The basic facts that are needed are the following: Let G be a locally compact abelian group. Then $\hat{G}$ is the collection of continuous unitary representations of G into C with the compact open topology as a function space on G and a group operation, $\oplus$, given by $(\gamma_1 \oplus \gamma_2)(g) = \gamma_1(g)\gamma_2(g), \gamma_1, \gamma_2 \in \hat{G}$, $g \in G$. That is, if $\gamma \in \hat{G}$ then γ is a continuous, complex-valued function on G, $|\gamma(g)| \equiv 1$, $\gamma(g_1 + g_2) = \gamma(g_1)\gamma(g_2), g, g_1, g_2 \in G$. The topology is given by defining neighborhoods of $\gamma_0 \in \hat{G}$ indexed by compact sets $C \subset G$ and $\epsilon > 0$, $N(\gamma_0, C, \epsilon) = \{\gamma \in \hat{G} : |\gamma(g) - \gamma_0(g)| < \epsilon$ for all $g \in C\}$.

The "big" theorem is that $\hat{G}$ is a locally compact abelian group and $(\hat{G})\hat{}$ is isomorphic to G. If $\hat{G}$ is isomorphic to G we say that G is <u>self dual</u>. An obvious corollary is that non-trivial groups have non-trivial duals and so we get the opening line of § 8 .

In §7 we use the notation consistently, but in §8 the notation tends to get overburdened at times and we will use K for K^+ as well $\hat{K}$ for $(K^+)\hat{}$.

The reason why little explicit attention is paid to the harmonic analysis of $\mathbb{R}^*$ is that, that analysis reduces to the study of two copies of the line. This follows from the fact that a multiplicative character on the units ($\pm$ 1) in $\mathbb{R}^*$ is either the identity character or is $x \to \mathrm{sgn}(x) = x/|x|$, $x \in \mathbb{R}^*$.

Chapter II: Fourier analysis on K, the one-dimension case

This chapter, in §1-§3, contains a treatment of the Fourier transform on K, as an additive group, which includes the theory of distributions on K. In §4 the Mellin transform (the Fourier transform on $K-\{0\}$ as a multiplicative group) is discussed and then in §5 the additive and multiplicative structures are meshed in a treatment of special functions on K. In §6 we treat a special topic; Fourier series on the ring of integers of K.

1. The L^1-theory

We assume, in general, that all functions are complex-valued and (Borel) measurable.

$L^p = L^p(K)$, $1 \le p \le \infty$, are the usual spaces, normed with $\|\cdot\|_p$.

$f \in L^p$, $1 \le p < \infty$ if $\|f\|_p = \left[\int_K |f(x)|^p dx\right]^{1/p} < \infty$.

$f \in L^\infty$ if $\|f\|_\infty = \text{ess sup}_{x \in K} |f(x)| < \infty$.

$\|f\|_p$ is called the L^p-norm of f. L^p is a Banach space if we identify, as usual, functions f and g such that $f(x) = g(x)$ a.e. C_0 is the class of continuous functions on K that vanish at infinity and is endowed with the L^∞-norm, $\|\cdot\|_\infty$.

<u>Examples</u>. Note that $\int_{|x|=q^k} 1dx = q^k(1-q^{-1})$. This follows from the observation that $\{x = q^k\} = \mathfrak{P}^{-k} \approx \mathfrak{P}^{-k-1}$ and that

$\mathfrak{P}^{-k} = \mathfrak{p}^{-k}\mathfrak{O}$, so $|\mathfrak{P}^{-k}| = |\mathfrak{p}^{-k}|\ |\mathfrak{O}| = q^k$. It is then easy to check that $\int_{|x|\leq 1} |x|^{\alpha} dx$ exists if $\alpha > -1$ and does not exist if $\alpha \leq -1$.

Similarly $\int_{|x|\leq 1} \ln \frac{1}{|x|}\, dx$ exists and $\int_{|x|\geq 1} |x|^{\alpha} dx$ exists if $\alpha < -1$ and does not exist if $\alpha \geq -1$.

__Definition.__ If $f \in L^1$ the Fourier transform of f is the function $\hat{f}$ defined by

$$\hat{f}(x) = \int_K f(\xi)\overline{\chi_x}(\xi)\, d\xi = \int_K f(\xi)\chi(-x\xi)\, d\xi .$$

__Observation.__ Since the restriction of a character to a fractional ideal $\mathfrak{P}^k$ is character on $\mathfrak{P}^k$, as a subgroup of K^+, we see that

$$\int_{|x|\leq q^k} \chi(x)\, dx = \begin{cases} q^k, & k \leq 0 \\ 0, & k > 0 \end{cases}, \text{ and so } \int_{|x| = q^k} \chi(x)\, dx = \begin{cases} q^k(1-q^{-1}), & k\leq 0 \\ -1, & k=1 \\ 0, & k>1 \end{cases}.$$

As a consequence we see that if $|x| = q^{\ell}$ then

$$\int_{|\xi| = q^k} \overline{\chi}_x(\xi)\, d\xi = \begin{cases} q^k(1-q^{-1}), & k \leq -\ell \ (|x| \leq q^{-k}) \\ -\frac{1}{|x|} = -q^{-\ell}, & k = -\ell+1 \ (|x| = q^{-k+1}) \\ 0, & k > -\ell+1 \ (|x| > q^{-k+1}) \end{cases}.$$

<u>Examples</u>. 1) Suppose $f(x) = |x|^{\alpha}$, $(\alpha > -1)$ $|x| \leq 1$ and is zero otherwise. Then

$$\hat{f}(x) = \begin{cases} \dfrac{1-q^{-1}}{1-q^{-(1+\alpha)}} & , \ |x| \leq 1 \\ \dfrac{1-q^{\alpha}}{1-q^{-(1+\alpha)}} \ \dfrac{1}{|x|^{1+\alpha}} & , \ |x| > 1 \end{cases}$$

2) Suppose $f(x) = \ln 1/|x|$, $|x| \leq 1$ is zero otherwise. Then

$$\hat{f}(x) = \begin{cases} \dfrac{\ln q}{1-q^{-1}} \, q^{-1} & , \ |x| \leq 1 \\ \dfrac{\ln q}{1-q^{-1}} \cdot \dfrac{1}{|x|} & , \ |x| > 1 \end{cases} .$$

<u>Theorem (1.1)</u>. (a) <u>The map</u> $f \to \hat{f}$ <u>is a bounded linear transformation of</u> L^1 <u>into</u> L^{∞}, $\|\hat{f}\|_{\infty} \leq \|f\|_1$.

(b) <u>If</u> $f \in L^1$ <u>then</u> $\hat{f}$ <u>is uniformly continuous</u>.

<u>Proof</u>. (a) Obvious. (b) $\hat{f}(x+h) - \hat{f}(x) = \int_K f(\xi)\overline{\chi}_x(\xi)[\overline{\chi}_h(\xi)-1]d\xi$.

The result then follows from the Lebesgue dominated convergence theorem using the fact that $\overline{\chi}_h$ is bounded and continuous as a function of h.

<u>Notation</u>. The <u>translation operator</u>, τ_h is defined for $h \in K$ by $(\tau_h f)(x) = f(x-h)$, whenever f is a function on K.

(1.2) <u>If</u> $f \in L^1$ <u>then</u> $(\tau_h f)^{\wedge} = \overline{\chi}_h \hat{f}$ <u>and</u> $(\chi_h f)^{\wedge} = \tau_h \hat{f}$.

<u>Notation</u>. (a) For $k \in \mathbb{Z}$ let Φ_k be the characteristic function of $\mathfrak{P}^k$. (b) A set of the form $h + \mathfrak{P}^k$ will be called a <u>sphere</u> with center h and radius q^{-k}.

Note that the characteristic function of $h + \mathfrak{P}^k = \tau_h\Phi_k$ and that $\tau_h\Phi_k$ is constant on cosets of $\mathfrak{P}^k$ (see (1.11).)

<u>Definition</u>. $\mathcal{S}$ is the space of finite linear combinations of functions of the form $\tau_h\Phi_k$, $h \in K$, $k \in \mathbb{Z}$.

<u>Proposition (1.3)</u>. *$\mathcal{S}$ is an algebra of continuous functions with compact support that separates points. Consequently $\mathcal{S}$ is dense in C_0 as well as in L^p, $1 \leq p < \infty$.*

<u>Proof</u>. All functions in $\mathcal{S}$ have compact support since each $\tau_h\Phi_k$ has that property. That the functions in $\mathcal{S}$ are continuous follows from the ultrametric inequality. Indeed if $y \in \mathfrak{P}^k$, then $\tau_h\Phi_k(x) = \tau_h\Phi_k(x+y)$ since $\tau_h\Phi_k$ is constant on cosets of $\mathfrak{P}^k$. That $\mathcal{S}$ is an algebra that separates points is left as a trivial exercise. The rest of the proposition follows from the Stone-Weierstrass theorem and the usual density arguments.

<u>Proposition (1.4)</u>. *For all* $k \in \mathbb{Z}$, $\hat{\Phi}_k = q^{-k}\Phi_{-k}$.

Proof. $\hat{\Phi}_k(x) = \int_K \Phi_k(\xi)\overline{\chi}_x(\xi)d\xi = \int_{\mathfrak{P}^k} \overline{\chi}_x(\xi)d\xi$. Since the restriction of a character on K^+ to $\mathfrak{P}^k$ is also a character we obtain

$$\hat{\Phi}_k(x) = \begin{cases} q^{-k}, & |x| \leq q^k \\ 0, & |x| > q^k \end{cases}.$$

In the first case $\overline{\chi}_x$ is trivial on $\mathfrak{P}^k$ and in the second it is non-trivial.

Notation. We define dilation operators. For $k \in \mathbb{Z}$ and f a function on K define $\delta_k f$ by $(\delta_k f)(x) = f(\mathfrak{p}^k x)$.

Remark. $\Phi_k = \delta_{-k}\Phi_0$.

(1.5) If $f \in L^1$ then $(\delta_k f)^\wedge = q^k(\delta_{-k}\hat{f})$.

Theorem (1.6) (Riemann-Lebesgue).

If $f \in L^1$ then $\hat{f}(x) \to 0$ as $|x| \to \infty$.

Proof. For $g \in \mathcal{S}$, $\hat{g}$ has compact support. Fix $\epsilon > 0$ and choose $g_\epsilon \in \mathcal{S}$ such that $\|f-g_\epsilon\|_1 < \epsilon$. For $x \notin \text{supp } \hat{g}_\epsilon$ we have $|\hat{f}(x)| = |(f-g_\epsilon)^\wedge(x)| \leq \|(f-g_\epsilon)^\wedge\|_\infty \leq \|f-g_\epsilon\|_1 < \epsilon$.

The Fourier transform can be extended to finite Borel measures μ as follows: $\hat{\mu}(x) = \int \overline{\chi}_x(\xi)d\mu(\xi)$.

Remark. (1.1) is valid for $\hat{\mu}$ (replace the L^1 norm of f by the total variation of μ), but (1.6) fails (if μ has mass 1 concentrated at the origin then $\hat{\mu} \equiv 1$).

We define convolution in the usual way. If f and g are functions, then $h = f * g$ is the function (when the defining integral exists):

$$h(x) = (f * g)(x) = \int f(x-z)g(z)dz = \int f(z)g(x-z)dz .$$

The following theorem is included for the sake of completeness. The proof is standard.

Theorem (1.7). If $f \in L^p$, $1 \leq p \leq \infty$ and $g \in L^1$ then $f * g \in L^p$ and $\|f * g\|_p \leq \|f\|_p \|g\|_1$.

For μ, a finite Borel measure, the convolution operator is extended as follows:

$$(\mu * f)(x) = \int f(x-z)d\mu(z) .$$

Remark. (1.7) extends to finite Borel measures μ in the sense that if μ is a finite Borel measure (with total variation $\|\mu\|_M$) and $f \in L^p$, $1 \leq p < \infty$ then $\mu * f \in L^p$ and $\|\mu * f\|_p \leq \|f\|_p \|\mu\|_M$. The case $p = \infty$ is replaced by: If $f \in C_0$ then $\mu * f \in C_0$ and $\|\mu * f\|_\infty \leq \|f\|_\infty \|\mu\|_M$.

In any case if f and g are in L^1 then $f * g \in L^1$ and the next result follows easily from the definitions:

Theorem (1.8). If $f, g \in L^1$ then $(f * g)^\wedge = \hat{f}\,\hat{g}$. If μ is a finite Borel measure then $(\mu * f)^\wedge = \hat{\mu}\,\hat{f}$.

We now consider the problem of inverting the Fourier transform. Formally we would expect: "$f(x) = \int \hat{f}(\xi)\chi_x(\xi)d\xi$" , but this does not always make sense since $\hat{f}$ is not necessarily in L^1 when $f \in L^1$. The example $f(x) = (\ell n\ 1/|x|)\Phi_0(x)$ discussed early, is a case in point. For $|x| > 1$, $\hat{f}(x) = (\ell n\ q/(1-q^{-1}))|x|^{-1}$.

Definition. If g is locally integrable and $k \in \mathbb{Z}$ let

$$A_k g = \int_K g\Phi_{-k} = \int_{|\xi| \leq q^k} g(\xi)d\xi \ .$$

If $g \in L^1$ it is clear that $A_k g \to \int g$ as $k \to \infty$. It is also clear that this limit may exist even though $\int g$ does not exist , as a Lebesgue integral.

Example. Let $\sum_{k=0}^{\infty} a_k$ be any convergent, not absolutely convergent series of complex numbers. Define g by the rule:

$$g(x) = \begin{cases} 0 & , |x| \leq 1 \\ a_k(1-q^{-1})q^{-k} & , |x| = q^k, \ k \geq 1 \end{cases} \ .$$

Then $A_k g \to \sum_{k=1}^{\infty} a_k$, but $\int |g| = \sum_{k=1}^{\infty} |a_k| = \infty$.

Theorem (1.9) (Multiplication Theorem). *If* $f,g \in L^1$ *then*
$\int \hat{f}(x)g(x)dx = \int f(x)\hat{g}(x)dx.$

Proof. From Fubini's Theorem and the definition of the Fourier transform,

$$\int \hat{f}(x)g(x)\,dx = \int\Big[\int f(\xi)\overline{\chi}_x(\xi)\,d\xi\Big]g(x)\,dx$$

$$= \int f(\xi)\Big[\int g(x)\overline{\chi}_\xi(x)\,dx\Big]d\xi = \int f(\xi)\hat{g}(\xi)\,d\xi \ .$$

Theorem (1.10). If $f \in L^1$, $k \in \mathbf{Z}$, then

$$A_k(\hat{f}\chi_x) = \int \hat{f}(\xi)\chi_x(\xi)\Phi_{-k}(\xi)\,d\xi$$

$$= q^k \int f(\xi)\Phi_k(x-\xi)\,d\xi = q^k \int_{|\xi-x| \le q^{-k}} f(\xi)\,d\xi \ .$$

Proof. Let $g = \chi_x\Phi_{-k}$. From (1.4) $(\Phi_{-k})^\wedge = q^k\Phi_k$. From (1.2) we have $\hat{g}(\xi) = (\chi_x\Phi_{-k})^\wedge(\xi) = \tau_x\hat{\Phi}_{-k}(\xi) = \hat{\Phi}_{-k}(\xi-x) = q^k\Phi_k(\xi-x) = q^k\Phi_k(x-\xi)$. An application of (1.9) completes the proof.

We now proceed to show that if $f \in L^1$ then $\{A_k(\hat{f}\chi_x)\}$ converges a.e. to $f(x)$, and then obtain the usual corollaries.

Lemma (1.11). Let S and T be two spheres in K. Then either S and T are disjoint or one sphere contains the other.

Proof. The result follows easily from the ultrametric inequality.

<u>Theorem (1.12) (A Covering Lemma in the style of N. Wiener).</u>

Let F *be a measurable subset of* K *of finite measure. Let* $\mathfrak{J} = \{T_\alpha\}$ *be a covering of* F *by spheres. Given* λ, $0 < \lambda < 1$, *there is a finite sub-collection* $\{T_k\}_{k=1}^N$ *of mutually disjoint sphere such that* $\sum_{k=1}^N |T_k| > \lambda |F|$.

Proof. We may assume that $\sup_\alpha |T_\alpha| < \infty$. Otherwise we choose a T_α such that $|T_\alpha| > |F|$ and we are done.

We define an equivalence relation on the element S of $\mathfrak{J}$. If $S, T \in \mathfrak{J}$ we say that $S \sim T$ iff $\exists\, V \in \mathfrak{J} \ni (S \subset V$ and $T \subset V)$.

Symmetry and reflexivity are obvious. To show transitivity, suppose $S \sim T$ and $T \sim R$. Then $\exists\, U, V \in \mathfrak{J} \ni S \subset U$, $T \subset U \cap V$, $R \subset V$. In particular $U \cap V = T \neq \emptyset$. This implies $U \subset V$ or $V \subset U$. Suppose $U \subset V$. Then we obtain $S \subset U \subset V$ so $S \sim R$. Else $V \subset U$, $R \subset V$ and $S \sim R$.

If we make the assumption that the elements of $\mathfrak{J}$ we distinct, which we may do, then $\{T_\alpha\}$ is a countable collection. In each equivalence class, there is a maximal element (the union of the element in that class) that is also in $\mathfrak{J}$, and maximal elements of distinct classes are disjoint. Thus there is a countable collection $\{T_k\}_{k=1}^\infty$ of mutually disjoint spheres in $\mathfrak{J}$ such that $F \subset \bigcup_{k=1}^\infty T_k$ and $|F| \leq |\bigcup_{k=1}^\infty T_k| = \sum_{k=1}^\infty |T_k|$. The result is now immediate for any $\lambda, 0<\lambda<$

Definition. If f is a locally integrable function, the Hardy-Littlewood maximal function of f is the function

$$Mf(x) = \sup_{k\in\mathbb{Z}} q^k \int_{|x-z|\leq q^{-k}} |f(z)|\,dz = \sup_{\substack{x\in S \\ S,\text{sphere}}} \frac{1}{|S|}\int_S |f| .$$

Theorem (1.13). *If* $f \in L^1$, $\lambda > 0$, *then* $|\{x : Mf(x) > \lambda\}| \leq \lambda^{-1}\|f\|_1$.

Proof. Let $E = \{x : Mf(x) > \lambda\}$, and F any measurable subset of E of finite measure. For $x \in F$ there is a sphere S_x such that $z \in S_x$ and $\int_{S_x} |f| > \lambda|S_x|$. The collection $\{S_x\}_{x\in F}$ satisfies the conditions of (1.12). Hence, given λ, $0 < \lambda < 1$ there is a finite sub-collection $\{S_k\}_{k=1}^N$ of mutually disjoint spheres such that $\sum_{k=1}^N |S_k| > \lambda|F|$. Thus, $\lambda|F| < \sum_{k=1}^N |S_k| < \sum_{k=1}^N \int_{S_k} |f| \leq \|f\|_1$. Thus, $|F| \leq \lambda^{-1}\|f\|_1$ and so $|E| \leq \lambda^{-1}\|f\|_1$.

Definition. Let f be locally integrable. Then $x \in K$ is said to be a regular point of f if

$$f_k(x) = q^k \int_{|x-z|\leq q^{-k}} f(z)\,dz \to f(x), \text{ as } k \to \infty .$$

Theorem (1.14). *If* f *is locally integrable then* a.e. $x \in K$ *is a regular point of* f.

<u>Proof</u>. We may assume that $f \in L^1$. (If $f \notin L^1$ replace it with $(\tau_x \Phi_0) f$.) For $g \in \mathscr{S}$, $g_k(x) = g(x)$ for large k. Fix $\epsilon > 0$ and choose $g \in \mathscr{S}$ such that $\|g - f\|_1 < \epsilon$. For large values of k we have that $f - f_k = (f - g) - (f - g)_k$, so that

$0 \leq \lim\sup |f(x) - f_k(x)| \leq |f(x) - g(x)| + \lim\sup |(f - g)_k(x)|$.

The result follows if for each $\delta > 0$, $E = \{x : \lim\sup |f(x) - f_k(x)| > \delta\}$ is of measure zero. But

$E \subset \{x : |f(x) - g(x)| > \delta/2\} \cup \{x : \lim\sup |(f-g)_k(x)| > \delta/2\} = E_1 \cup E_2$.

$|E_1| \leq 2\delta^{-1} \|f - g\|_1$, $|E_2| \leq 2\delta^{-1} \|f - g\|_1$ (the second of these by (1.13) so $|E| \leq 4\delta \|f - g\|_1 = 4\delta\epsilon$. Let $\epsilon \to 0$ and we see that $|E| = 0$.

<u>Corollary (1.15)</u>. <u>If</u> $f \in L^1$ <u>then</u> $A_k(\chi_x \hat{f}) \to f(x)$ a.e.. <u>In particular, it converges at each point of continuity of</u> f.

<u>Proof</u>. An immediate corollary of (1.10) and (1.14).

<u>Corollary (1.16)</u>. <u>If</u> f <u>and</u> $\hat{f}$ <u>are both integrable then</u> f <u>is equal, a.e., to a continuous function</u>. <u>With</u> f <u>modified (on a set of measure zero) to be continuous we obtain</u>

$$f(x) = \int \hat{f}(\xi) \chi_x(\xi) d\xi \quad \text{for all } x.$$

<u>Proof</u>. If $\hat{f}$ is integrable then $A_k(\chi_x \hat{f})$ converges to a continuous function; namely $\int \hat{f}\,\chi_x$ (1.1(b)). By (1.15) we see that f is continuous a.e. Modify f and a set of measure zero and we are done.

<u>Corollary (1.17)</u>. *If* $f, g \in L^1$ *and* $\hat{f} = \hat{g}$ *then* $f(x) = g(x)$ *a.e.*

<u>Proof</u>. $(f-g)^{\wedge} \equiv 0$. By (1.16) $(f-g)(x) = 0$ a.e.

<u>Corollary (1.18)</u>. *If* $f \in L^1$, $\hat{f} \geq 0$ *and* f *is continuous at* 0, *then* $\hat{f} \in L^1$ *and* $f(x) = \int \hat{f}(\xi)\chi_x(\xi)\,d\xi$ *at each regular point of* f. *In particular*, $f(0) = \int \hat{f}(\xi)\,d\xi$.

<u>Proof</u>. We only need to show that $\hat{f} \in L^1$. Since f is continuous at 0 we obtain (from (1.10) and (1.14)) that

$f(0) = \lim_{k\to\infty} \int_{|x| \leq q^k} \hat{f}(\xi)\,d\xi$. Since $\hat{f} \geq 0$, Fatou's Lemma shows that $\hat{f} \in L^1$.

<u>Notation</u>. For $k, \ell \in \mathbb{Z}$, $k \vee \ell = \max[k,\ell]$, $k \wedge \ell = \min[k,\ell]$.

<u>Definition</u>. For $k \in \mathbb{Z}$ let $R(x,k) = q^{-k}\Phi_{-k}$.

(1.19) $\quad \hat{R}(\cdot,k) = \Phi_k, \ \hat{\Phi}_k = R(\cdot,k)$

(1.20) $\quad R(\cdot,k)*R(\cdot,\ell) = R(\cdot,k\vee\ell)$

(1.19) is a restatement of (1.4). From (1.8) and (1.19) we see that $(R(\cdot,k)*R(\cdot,\ell))^{\wedge} = \Phi_k\Phi_\ell = \Phi_{k\vee\ell} = \hat{R}(\cdot,k\vee\ell)$. An application of (1.17) completes the proof.

2. The L^2-theory

<u>Theorem (2.1)</u>. <u>If</u> $f \in L^1 \cap L^2$ <u>then</u> $\|\hat{f}\|_2 = \|f\|_2$.

<u>Proof</u>. Let $g(x) = \overline{f}(-x)$. Then $\hat{g} = \overline{\hat{f}}$. Since $f,g \in L^1$, $f*g \in L^1$, $(f*g)^{\wedge} = \hat{f}\hat{g} = |\hat{f}|^2 \geq 0$. Since $f,g \in L^2$, $f*g$ is continuous. To see this we write

$$|(f*g)(x+y) - f*g(x)| = \left|\int (f(x+y-z) - f(x-z))g(z)dz\right|$$

$$< \left[\int |f(x+y-z) - f(x-z)|^2 dz\right]^{\frac{1}{2}}\|g\|_2 = \|f(\cdot+y) - f(\cdot)\|_2 \, \|g\|_2 \, .$$

But $\|f(\cdot+y) - f(\cdot)\|_2 = o(1)$ as $y \to 0$.

We then apply (1.18) and we see that $(f*g)^{\wedge} = |\hat{f}|^2 \in L^1$ and $\int|\hat{f}|^2 = (f*g)(0) = \int g(-z)f(z)dz = \int \overline{f}f = \int|f|^2$.

From this theorem we see that the map $f \to \hat{f}$ is an L^2-isometry on $L^1 \cap L^2$, which is a dense subspace of L^2. We now extend the

Fourier transform to L^2. From (2.1) it is easily seen to an isometry on L^2, and agrees with the Fourier transform on L^1 on the set $L^1 \cap L^2$.

<u>Definition</u>. If $f \in L^2$, let $f_k = f\Phi_{-k}$ and

$$\hat{f}(x) = \lim_{k\to\infty} (f_k)^{\wedge}(x) = \lim_{|\xi|\leq q^k} \int f(\xi)\bar{\chi}_x(\xi)\,d\xi,$$

where the limit is taken in L^2.

<u>Theorem (2.2) (Multiplication)</u>. <u>If</u> $f, g \in L^2$ <u>then</u> $\int f\hat{g} = \int \hat{f} g$.

<u>Proof</u>. $\{f_k\} \to f$ and $\{g_k\} \to g$ in L^2. Thus, $\{\hat{f}_k\} \to \hat{f}$, $\{\hat{g}_k\} \to \hat{g}$ in L^2. By (2.1) we see that $\int f_k\hat{g}_k = \int \hat{f}_k g_k$ for each k. We use Schwartz's inequality and see that the left hand side converges to $\int f\hat{g}$ and the right hand side of $\int \hat{f} g$.

<u>Remark</u>. The relations (1.2) and (1.5) hold for $f \in L^2$.

<u>Theorem (2.3)</u>. <u>The Fourier transform is unitary on</u> L^2.

<u>Proof</u>. $f \to \hat{f}$ is a linear isometry. We only need to show that it is onto. If not there is a $g \in L^2$ such that $\int \hat{f} g = 0$ for all $f \in L^2$, $\|g\|_2 \neq 0$. But (2.2) then implies that $\int f\hat{g} = 0$ for all $f \in L^2$. This implies that $\hat{g} \equiv 0$. But $\|\hat{g}\|_2 = \|g\| \neq 0$, a contradiction.

Notation. The inverse mapping to the Fourier transform is denoted $f \to f^{\vee}$. We denote the reflection operator for a function f on K by $\tilde{f}$ $(\tilde{f}(x) = f(-x))$.

In the proof of (2.1) we noted that $(\bar{\tilde{f}})^{\wedge} = \bar{\hat{f}}$. This extends to L^2 as usual. Similarly $(\bar{f})^{\wedge} = \overline{(\tilde{f})^{\wedge}}$. Thus if $f,g \in L^2$, $\int \hat{f}\,\bar{g} = \int f\,(\bar{g})^{\wedge} = \int f\,\overline{(\tilde{g})^{\wedge}}$ and the map $g \to (\tilde{g})^{\wedge}$ is adjoint to $f \to \hat{f}$. But $f \to \hat{f}$ is unitary so the adjoint of $f \to \hat{f}$ is $g \to g^{\vee}$.

(2.4) *If* $f \in L^2$, $f^{\vee} = (\tilde{f})^{\wedge}$.

Since $f \to \hat{f}$ is unitary we also have,

(2.5) *If* $f,g \in L^2$ *then* $\int f\,\bar{g} = \int \hat{f}\,\bar{\hat{g}}$,

which is also a rewriting of (2.2) using (2.4).

Theorem (2.6). *If* $f \in L^2$ *then* $\lim_{k\to\infty} \int_{|\xi| \le q^k} \bar{\chi}_x(\xi) f(\xi) = \hat{f}(x)$ a.e.

Proof. We use (2.2), the relations (1.2) and the fact that Φ_k, f and $\hat{f}$ are in L^2 and argue as in (1.8):

$$\int_{|\xi|\le q^k} f(\xi)\bar{\chi}_x(\xi)\,d\xi = \int f\,(\overline{\Phi_{-k}\chi_x}) = \int \hat{f}\,(\overline{\Phi_{-k}\chi_x)^{\wedge}}$$

$$= \int \hat{f}\,\overline{\tau_x(\Phi_{-k})^{\wedge}} = q^k \int \hat{f}\,\overline{\tau_x\Phi_k} = q^k \int f(\xi)\Phi_k(\xi - x)\,d\xi$$

$$= q^k \int_{|\xi - x| \le q^{-k}} \hat{f}(\xi)\,d\xi \ .$$

$f \in L^2 \Rightarrow \hat{f} \in L^2$. Thus $\hat{f}$ is locally integrable and an application of (1.14) completes the proof.

At this point we see that the Fourier transform can be extended to $L^1 + L^2$ (which is to say, any function $f = f_1 + f_2$ with $f_1 \in L^1$, $f_2 \in L^2$). That is, for such an f we set $\hat{f} = \hat{f}_1 + \hat{f}_2$ where the Fourier transform is defined on L^1 and L^2 respectively. It is easy to check that it is well defined as a locally integrable function. In particular the Fourier transform is defined for $f \in L^p$, $1 \leq p \leq 2$.

<u>Theorem (2.7)</u>. <u>If</u> $f \in L^1$, $g \in L^p$, $1 \leq p \leq 2$, <u>then</u> $(f * g)\hat{} = \hat{f}\,\hat{g}$ a.e.

<u>Proof</u>. From (1.8) the result holds for $p = 1$. It will suffice to establish the result for $p = 2$, using the linearity of the Fourier transform and the fact that $f * (g_1 + g_2) = f * g_1 + f * g_2$.

Suppose $g \in L^2$. From (1.8) and (2.6) we have $(f * (\Phi_{-k} g))\hat{} = \hat{f}(\Phi_{-k} g)\hat{} \rightarrow \hat{f}\,\hat{g}$ a.e. From (1.7) we see that $h \rightarrow f * h$ is continuous in $L^2 (f \in L^1)$, so $f * (\Phi_{-k} g) \rightarrow f * g$ in L^2. Since the Fourier transform is continuous in L^2, $(f * (\Phi_{-k} g))\hat{} \rightarrow (f * g)\hat{}$ in L^2. Thus $(f * g)\hat{} = \hat{f}\,\hat{g}$ a.e.

3. Distributions on K

Let $\mathscr{S}$ be the class of functions defined in §1. This class can be described alternatively as follows:

(3.1) $\varphi \in \mathscr{S}$ *iff there are integers* k, ℓ *such that* φ *is constant on cosets of* $\mathfrak{P}^k$ *and is supported* $\mathfrak{P}^\ell$.

The following result is crucial and is the local field version of the usual situation where the "smoothness of the function" reflects the "rate of decrease at infinity" of its Fourier transform and conversely.

Theorem (3.2). *If* $\varphi \in \mathscr{S}$ *is constant on cosets of* $\mathfrak{P}^k$ *and is supported on* $\mathfrak{P}^\ell$ *then* $\hat{\varphi} \in \mathscr{S}$ *is constant on cosets of* $\mathfrak{P}^{-\ell}$ *and is supported on* $\mathfrak{P}^{-k}$.

Proof. We may assume that $k \geq \ell$ and that φ is a finite linear combination of functions of the form $\tau_h \Phi_k$, with $h \in \mathfrak{P}^\ell$. But $(\tau_h \Phi_k)^\wedge = q^{-k} \chi_h \Phi_{-k}$ which is constant on cosets of $\mathfrak{P}^{-\ell}$ and is supported on $\mathfrak{P}^{-k}$, and hence so is $\hat{\varphi}$.

Theorem (3.3). *If* $\varphi, \psi \in \mathscr{S}$, $h \in K$ *then* $\tau_h \varphi$, $\tilde{\varphi}$, $\varphi * \psi \in \mathscr{S}$.

Proof. The result for $\tau_h \varphi$ is a triviality, as in the result for $\tilde{\varphi}$. Clearly $\mathscr{S}$ is closed under multiplication of functions. By (1.7) we have $(\varphi * \psi)^{\wedge} = \hat{\varphi}\,\hat{\psi} \in \mathcal{S}$. But (3.2), (1.16) and (2.4) show that $(\varphi \in \mathcal{S}$ iff $\hat{\varphi} \in \mathcal{S})$ so $\varphi * \psi \in \mathcal{S}$.

Remark. To clarify the last part of the proof of (3.3); (3.2), (1.16) and (2.4) show that if we define the maps $\varphi \to \hat{\varphi}$, $\varphi \to \check{\varphi}$ by $\hat{\varphi}(x) = \int \varphi(\xi)\overline{\chi_x}(\xi)d\xi$, $\check{\varphi}(x) = \int \varphi(\xi)\chi_x(\xi)d\xi$, $\varphi \in \mathcal{S}$, then they each map $\mathcal{S}$ onto $\mathcal{S}$ and are inverses to each other.

$\mathcal{S}$ is provided with a topology as a topological vector space as follows: We define a null sequence in $\mathcal{S}$ as a sequence $\{\varphi_n\}$ of functions in $\mathcal{S}$ such that there is a fixed pair of integers k and ℓ such that each φ_n is constant on cosets of $\mathfrak{P}^k$ and is supported on $\mathfrak{P}^\ell$ and the sequence tends (uniformly) to zero. $\mathcal{S}$ is referred to as the <u>space of testing functions</u>.

(3.4) <u>$\mathcal{S}$ is complete.</u>

Easy.

(3.5) <u>$\mathcal{S}$ is separable.</u>

The set of rationally-valued elements of $\mathcal{S}$ is countable and is also dense in $\mathcal{S}$.

(3.6) <u>The Fourier transform is a homeomorphism of $\mathcal{S}$ onto $\mathcal{S}$.</u>

For "onto" see the remark following (3.3). Continuity follows from (1.1(a)).

(3.7) $\varphi \to \tilde{\varphi}$ <u>and</u> $\varphi \to \tau_h \varphi (h \in K)$ <u>are homeomorphisms of</u> $\mathscr{S}$ <u>onto</u> $\mathscr{S}$.

Obvious.

(3.8) $\mathscr{S}$ <u>is continuously contained and dense</u> in L^p, $1 \le p < \infty$ <u>and in</u> C_0.

Density was discussed at (1.3). For continuous containment note that $(\varphi_k \xrightarrow{\mathscr{S}} \varphi) \Rightarrow (\varphi_k \xrightarrow{L^\infty} \varphi)$ from the definition of the topology of $\mathscr{S}$. For L^p, $1 \le p < \infty$ we suppose that each φ_k is supported on $\mathfrak{P}^{\ell}$ (and hence so is φ), then
$\|\varphi_k - \varphi\|_p \le q^{-\ell/p} \|\varphi - \varphi_k\|_\infty = o(1)$, as $k \to \infty$.

The collection, $\mathscr{S}'$, of continuous linear functionals on $\mathscr{S}$ is called the <u>space of distributions</u> $\mathscr{S}'$ is given the Weak* topology. The action of $f \in \mathscr{S}'$ on $\varphi \in \mathscr{S}$ is denote (f,φ). $f \in \mathscr{S}'$, $\varphi_k \xrightarrow{\mathscr{S}} \varphi$ implies $(f,\varphi_k) \to (f,\varphi)$.

We consider some examples:

(1) Suppose f is locally integrable. Then $(f,\varphi) = \int f\varphi$ defines an element of $\mathscr{S}'$. To see this note that if $\varphi_k \to \varphi(\mathscr{S})$, then $\sup_k |\varphi_k| = \psi \in \mathscr{S}$ and the Dominated Convergence theorem applies. If, in particular, $f \in L^p$, $1 \le p \le \infty$, then f induces an element of $\mathscr{S}'$.

If f is a distribution and g is a locally integrable function such that $(f,\varphi) = \int g\varphi$ for all $\varphi \in \mathcal{S}$ we say that <u>f is a function</u> and that <u>f is the function g</u>.

(2) If μ is a Borel measure, not necessarily finite, then $(\mu,\varphi) = \int \varphi \, d\mu$ defines a distribution. (Convergence in $\mathcal{S}$ implies uniform convergence on a compact support set.) In a similar fashion we use the expression <u>$f \in \mathcal{S}'$ is a measure</u> and <u>f is the measure μ</u> when there is a Borel measure μ such that $(f,\varphi) = \int \varphi \, d\mu$ for all $\varphi \in \mathcal{S}$.

(3) An important example of a distribution which is a measure is the evaluation functional: $(\delta,\varphi) = \varphi(0)$. δ, as a measure, is the Dirac delta which assigns mass 1 to the origin.

(4) An example of (1) is: $(1,\varphi) = \int \varphi$ which is the function that is identically 1.

Our definition of the Fourier transform is motivated by the multiplication formula. Recall that if $f,g \in L^1$ or $f,g \in L^2$ then $\int f\hat{g} = \int \hat{f} g$ so that if $g \in \mathcal{S}$ and $f \in L^1 + L^2$ we have $(f,\hat{g}) = (\hat{f},g)$, by a simple calculation.

<u>Definition</u>. If $f \in \mathcal{S}'$ then $\hat{f}$, called the <u>Fourier transform</u> of f, is defined by: $(\hat{f},\varphi) = (f,\hat{\varphi})$ for all $\varphi \in \mathcal{S}$.

<u>Remark</u>. Since $\varphi \to \hat{\varphi}$ is a homeomorphism of $\mathcal{S}$ we see that $\hat{f} \in \mathcal{S}'$.

Examples: Find $\hat{\delta}$: $(\hat{\delta},\varphi) = (\delta,\hat{\varphi}) = \hat{\varphi}(0) = \int \varphi$. Thus $\hat{\delta} = 1$.

Find $\hat{1}$: $(\hat{1},\varphi) = (1,\hat{\varphi}) = \int \hat{\varphi} = \varphi(0)$, $\hat{1} = \delta$.

Clearly, this definition agrees with our definition of $\hat{f}$ on L^1 and L^2 and hence on $L^1 + L^2$. A substantial improvement is that we may now define $\hat{f}$ for $f \in L^p$, $2 < p \leq \infty$.

Definition. The inverse Fourier transform of $f \in \mathscr{S}'$ is defined by $(f^\vee,\varphi) = (f,\varphi^\vee)$ for all $\varphi \in \mathscr{S}$.

Remark. If $f \in \mathscr{S}'$, then $f^\vee \in \mathscr{S}'$.

Proposition (3.9). If $f \in \mathscr{S}'$ then $(\hat{f})^\vee = f = (f^\vee)^\wedge$.

These relations follow from the corresponding facts for $\varphi \in \mathscr{S}$ and the definition of $\hat{f}$ and $f^\vee$.

Theorem (3.10). The map $f \rightarrow \hat{f}$ is a homeomorphism of $\mathscr{S}'$ onto $\mathscr{S}'$

Proof. For onto see (3.9). Continuity from the observation that for a fixed $\varphi \in \mathscr{S}$ the set $\{f \in \mathscr{S}': |(f,\hat{\varphi})| < \varepsilon\}$ is mapped by the Fourier transform onto $\{\hat{f} \in \mathscr{S}': |(\hat{f},\varphi)| < \varepsilon\}$.

Definition. For $f \in \mathscr{S}'$, $h \in K$ define $\tilde{f}$ and $\tau_h f$, as elements in $\mathscr{S}'$ by : $(\tilde{f},\varphi) = (f,\tilde{\varphi})$ and $(\tau_h f,\varphi) = (f,\tau_{-h}\varphi)$ for all $\varphi \in \mathscr{S}$.

Proposition (3.11). *For* $h \in K$, $f \to \tau_h f$ *and* $f \to \tilde{f}$ *are homeomorphisms of* $\mathcal{S}'$ *onto* $\mathcal{S}'$.

4. The Mellin transform-Fourier analysis on K^*

For $x \in K^*$ let $x = \mathfrak{p}^k x'$ with $x' \in \mathfrak{O}^*$, $k \in \mathbb{Z}$.

For $\pi \in \hat{K}^*$ let $\pi = \pi^* |\cdot|^{i\alpha}$, $\pi^* \in \hat{\mathfrak{O}}^*$, $-\pi/\ln q < \alpha \leq \pi/\ln q$.

(4.1) For $f \in L^1(K^*, dx^*)$,

$$\int_{K^*} f(x)\,dx^* = \int_K f(x) \frac{dx}{|x|} = \sum_{k=-\infty}^{+\infty} \int_{x' \in \mathfrak{O}^*} f(\mathfrak{p}^k x')\,dx'$$

(4.2) For $g \in L^1(\hat{K}^*, d\pi)$,

$$\int_{\hat{K}^*} g(\pi)\,d\pi = \sum_{\pi^* \in \hat{\mathfrak{O}}^*} \frac{1}{a} \int_{-\pi/\ln q}^{\pi/\ln q} g(\pi^* |\cdot|^{i\alpha})\,d\alpha$$

where a is a positive constant that we will determine later.

The collection of *test functions* $\mathcal{S}^*$, *on* K^* is the set of complex-valued functions on K^* with compact support ($\varphi \in \mathcal{S}^* \Rightarrow \varphi$ is supported on $\{q^{-n} \leq |x| \leq q^n\}$ for some n) and are constant on cosets (in K^*) of some A_m. A topology on $\mathcal{S}^*$ (as a topological vector space) by defining $\{\varphi_k\}$ to be a null sequence if each $\varphi_k \in \mathcal{S}^*$, the φ_k have common compact support, are all constant on cosets of the same subgroup A_m and tend uniformly to zero. The topological dual, $\mathscr{S}^{*\prime}$ of $\mathscr{S}^*$ is called the <u>space of distributions on K^*</u>. $\mathscr{S}^{*\prime}$ is supplied with the Weak*topology.

<u>Lemma (4.3)</u>. <u>Suppose</u> $\varphi \in \mathscr{S}$ <u>and</u> φ <u>is constant on cosets (in</u> K^+) <u>of</u> $\mathfrak{P}^n$, <u>then</u> φ <u>is constant on cosets (in</u> K^*) <u>of</u> A_{n-k} <u>contained in</u> $\mathfrak{P}^k$ <u>for all</u> $k \leq n$. <u>In particular, if</u> $\varphi \in \mathscr{S}$ <u>and</u> $\varphi(0) = 0$ <u>then</u> $\varphi \in \mathscr{S}^*$.

<u>Proof</u>. Suppose $a A_{n-k} = b A_{n-k}$, $|a| = |b| \leq q^{-k}$, $k \leq n$. Then $ab^{-1} \in A_{n-k}$ so $1 - ab^{-1} \in \mathfrak{P}^{n-k}$. Then $b - a = b(1 - ab^{-1})$ $\in \mathfrak{P}^k \mathfrak{P}^{n-k} \subset \mathfrak{P}^n \Longrightarrow \varphi(a) = \varphi(b)$.

Now suppose φ is supported on $\mathfrak{P}^m$ where we may assume that $m \leq n$. Then φ is constant on cosets (in K^*) of A_{n-m}. If $\varphi(0) = 0$ (so $\varphi(x) \equiv 0$ on $\mathfrak{P}^n$) φ has compact support on K^* so $\varphi \in \mathscr{S}^*$.

<u>Lemma (4.4)</u>. <u>Suppose</u> $\varphi \in \mathscr{S}^{*\prime}$, φ <u>constant on cosets (in</u> K^*) <u>of</u> A_n, $n \geq 1$. <u>Then</u> φ <u>is constant on cosets (in</u> K^+) <u>of</u> $\mathfrak{P}^{n+k}$ <u>contained in</u> $K \sim \mathfrak{P}^{k+1}$ <u>for all</u> $k \in \mathbb{Z}$. <u>In particular if</u> $\varphi \in \mathscr{S}^*$ <u>then</u> $\varphi \in \mathscr{S}$.

<u>Proof</u>. Left as an easy exercise.

If π is ramified of degree h, $h \geq 1$, let $\deg(\pi) = \deg(\pi^*) = h$. If π is unramified let $\deg(\pi) = \deg(\pi^*) = 0$.

The collection of <u>test functions, $\hat{\mathscr{S}}^*$, on $\hat{K}^*$</u> is the set of complex-valued functions on $\hat{K}^*$ with compact support (i.e., $g(\pi)$ is supported on a set $\deg(\pi) \leq h$ for some h) and the restriction to each circle $\mathbb{T}_{\pi^*}$ is a "trigonometric polynomial". That is, there are integers m, $h \geq 0$ such that if $\varphi \in \hat{\mathscr{S}}^*$ then $\varphi(\pi) = \varphi(\pi^*|\cdot|^{i\alpha}) = \sum_{\nu=-m}^{m} a_\nu(\pi^*)q^{i\nu\alpha}$ and $a_\nu(\pi^*) = 0$ if $\deg(\pi^*) > h$. $\hat{\mathscr{S}}^*$ is given the topology induced by defining a null sequence $\{\varphi_h\}$ to be a sequence in $\hat{\mathscr{S}}^*$ with a uniform choice of m and h, that tends uniformly to zero. The <u>space of distributions</u>, $\hat{\mathscr{S}}^{*\prime}$ is the topological dual of $\hat{\mathscr{S}}^*$ with the Weak*topology. For $f \in \mathscr{S}^{*\prime}$ (or $\hat{\mathscr{S}}^{*\prime}$) and $\varphi \in \mathscr{S}^*$ (or $\hat{\mathscr{S}}^*$) the action of f on φ is denoted $< f,\varphi >$. The context will resolve the ambiguity.

Let Ψ_n be the characteristic function of A_n. Let Λ_n be the characteristic function of $A_n' = \{\mathbb{T}_{\pi^*}: \deg(\pi^*) \leq n\}$.

The Fourier transform of K^* is called the <u>Mellin transform</u>. For $f \in L^1(K^*)$ the Mellin transform of f is $\mathfrak{M}f(\pi)$, $\pi \in \hat{K}^*$,

$$\mathfrak{M}f(\pi) = \int f(x)\pi(x)\frac{dx}{|x|} = \sum_{k=-\infty}^{+\infty} q^{ik\alpha}\int_{x'\in\mathfrak{D}^*} f(\mathfrak{p}^{-k}x')\pi^*(x')dx' .$$

For $g \in L^1(\hat{K}^*)$ the <u>Inverse Mellin transform</u> of g is $\mathfrak{M}^{-1}g(x)$, $x \in K^*$,

$$\mathfrak{M}^{-1}g(x) = \int g(\pi)\pi^{-1}(x)d\pi = \sum_{\pi^*\in\hat{\mathfrak{D}}^*} \pi^*(x')\, \frac{1}{a}\int_{-\pi/\ln q}^{\pi/\ln q} g(\pi^*|\cdot|^{i\alpha})|x|^{i\alpha}d\alpha\ .$$

These transforms are then extended to L^2 in the usual way and a is chosen so Plancherel holds. That is, so that $\|f\|_2 = \|\mathfrak{M} f\|_2$, $\|g\|_2 = \|\mathfrak{M}^{-1}g\|_2$ and $\mathfrak{M}^{-1}\mathfrak{M} = \mathfrak{M}\,\mathfrak{M}^{-1} = \mathrm{Id}$.

<u>Lemma (4.5)</u>. <u>If</u> $\varphi \in \mathcal{S}^*$ <u>then</u> φ <u>is supported on</u> $\{q^{-n} \le |x| \le q^n\}$ <u>and is constant on cosets (in</u> K^*<u>) of</u> A_m <u>iff</u> $\mathfrak{M}\varphi \in \hat{\mathcal{S}}^*$, $\mathfrak{M}\varphi$ <u>is supported on</u> A'_m <u>and the restrictions of</u> $\mathfrak{M}\varphi$ <u>to</u> $\mathbb{T}_{\pi^*}$ <u>are of degree bounded by</u> n.

<u>Proof</u>. This is the multiplicative version of (3.2). We leave the proof as an exercise.

The Mellin transform and its inverse are then extended to $\mathcal{S}^{*\prime}$ and $\hat{\mathcal{S}}^{*\prime}$, with $\mathfrak{M}:\mathcal{S}^{*\prime} \to \hat{\mathcal{S}}^{*\prime}$ and $\mathfrak{M}^{-1}: \hat{\mathcal{S}}^{*\prime} \to \mathcal{S}^{*\prime}$ as homeomorphisms. For $f \in \mathcal{S}^{*\prime}$, $\psi \in \hat{\mathcal{S}}^*$, $<\mathfrak{M}f,\psi> = <f,(\mathfrak{M}^{-1}\psi)^{\sim}>$. For $g \in \hat{\mathcal{S}}^{*\prime}$ and $\varphi \in \mathcal{S}^*$, $<\mathfrak{M}^{-1}g,\varphi> = <g,(\mathfrak{M}\varphi)^{\sim}>$, where the "$\sim$"- operation is reflection in the appropriate group structure. That is, for φ a function on K^*, $\tilde{\varphi}(x) = \varphi(x^{-1})$, and for ψ a function on $\hat{K}^*$, $\tilde{\psi}(\pi) = \psi(\pi^{-1})$.

We now determine the correct value for the constant a.

$$\mathfrak{M}\Psi_n(\pi) = \int_{A_n} \pi(x)\,\frac{dx}{|x|} = \int_{A_n} \pi^*(x')dx' = \begin{cases} |A_n|\,, & \pi \in A'_n \\ 0\,, & \pi \notin A'_n \end{cases}\ .$$

Thus, $(1-q^{-1})^{-1}\mathfrak{M}\Psi_0 = \Lambda_0$, $q^n\mathfrak{M}\Psi_n = \Lambda_n$, $n \geq 1$. From Plancherel it follows that

$$|A_0'| = \|\Lambda_0\|_2^2 = (1-q^{-1})^{-2}\,\|\Psi_0\|_2^2 = (1-q^{-1})^{-2}\,|A_0| = (1-q^{-1})^{-1} .$$

Using the notation above for test functions on $\hat{K}^*$ we see that Λ_0 is the test function $\{a_\nu(\pi^*)\}$ where $a_\nu(\pi^*) = 1$ if $\nu = 0$ and $\deg(\pi^*) = 0$ and is zero otherwise. Thus,

$$(1-q^{-1})^{-1} = |A_0'| = \int \Lambda_0 \, d\pi = \frac{1}{a}\int_{-\pi/\ln q}^{\pi/\ln q} 1 \, d\alpha = \frac{2\pi}{a \ln q} . \quad \text{Thus,}$$

$a = 2\pi(1-q^{-1})/\ln q$.

Several other facts are now easily seen:

(4.6) $\{(1-q^{-1})\Psi_0 , \; q^n \Psi_n\}_{n=1}^{\infty}$ <u>is a non-negative approximation to the identity on</u> K^* <u>with all its terms in</u> $\mathscr{S}^*$.

The number of (unitary) characters on $\mathfrak{D}^*$ that are in A_m' is 1 if $m = 0$ and is $q^m(1-q^{-1})$ for $m \geq 1$. Thus, there is one unramified character on $\mathfrak{D}^*$, $(q-1)-1 = q-2$ characters that are ramified of degree 1 and $q^m(1-q^{-1}) - q^{n-1}(1-q^{-1}) = q^{m-2}(q-1)^2$ characters that are ramified of degree m, $m \geq 2$.

For functions f, that are locally integrable on K^* (i.e., f is integrable on each set $\{q^{-n} \leq |x| \leq q^n\}$, $n \geq 0$) we define two methods of summability of integrals on K.

$$\text{(4.7)} \qquad P\int f(x)dx = P\int f = \lim_{n\to\infty} \int_{q^{-n}\leq|x|\leq q^n} f(x)dx$$

$$\text{(4.8)} \qquad (C,1)\int f(x)dx = (C,1)\int f = \lim_{n\to\infty} \frac{1}{n+1} \sum_{\nu=0}^{n} \int_{q^{-\nu}\leq|x|\leq q^{\nu}} f(x)dx$$

If $P\int f$ exists so does $(C,1)\int f$ and they are equal. If $\int f$ exists, then $P\int f$ exists and $\int f = P\int f$. These are standard facts about summability.

The notion of a distribution being a function is extended to include those cases where f is a locally integrable function on K^* and $\varphi \to P\int f\varphi$ or $\varphi \to (C,1)\int f\varphi$ defines a distribution (for $\varphi \in \mathscr{S}$). In such a case the distribution is denoted Pf or $(C,1)f$ and will be written: (Pf,φ), $((C,1)f,\varphi)$.

As we shall see in the next section if π^* is a ramified unitary character on $\mathfrak{D}^*$ and $\operatorname{Re}(\alpha) = -1$, then $P\,\pi^*(x)|x|^{\alpha} \in \mathscr{S}'$. Similarly if $\operatorname{Re}(\alpha) = -1$, $\alpha \neq -1$ then $(C,1)|x|^{\alpha} \in \mathscr{S}'$.

We now describe these methods of summability in terms of distributions on K^* and $\hat{K}^*$.

Consider the Dirichelet kernel:

$$D_n(x) = \frac{1}{2} + \sum_{\nu=1}^{n} \cos \nu x = \frac{1}{2}\sum_{|\nu|\le n} e^{i\nu x} = \frac{\sin((n+1/2)x)}{2\sin(x/2)} .$$

Let $d_n(\pi) = d_n(\pi^*|\cdot|^{i\alpha}) = \begin{cases} 2(1-q^{-1})\ D_n(\alpha\ \ln q), & \deg(\pi^*) = 0 \\ 0 & , \deg(\pi^*) \neq 0 \end{cases}$.

Then $d_n \in \hat{\mathscr{S}}^*$ and $d_n(|\cdot|^{i\alpha}) = (1-q^{-1}) \sum_{\nu=-n}^{n} q^{i\nu\alpha}$.

Consider the Fejer kernel:

$$K_n(x) = \frac{1}{n+1}\sum_{\nu=0}^{n} D_\nu(x) = \frac{2}{2n+1}\left[\frac{\sin((n+1)x/2)}{2\sin(x/2)}\right]^2 = \frac{1}{2}\sum_{|\nu|\le n}\left(1-\frac{|\nu|}{n+1}\right)e^{i\nu x}.$$

Let $k_n(\pi) = k_n(\pi^*|\cdot|^{i\alpha}) = \begin{cases} 2(1-q^{-1})\ K_n(\alpha\ \ln q), & \deg(\pi^*) = 0 \\ 0 & , \deg(\pi^*) \neq 0 \end{cases}$.

Then $k_n \in \hat{\mathscr{S}}^*$ and $k_n(|\cdot|^{i\alpha}) = (1-q^{-1}) \sum_{\nu=-n}^{n} \left(1-\frac{|\nu|}{n+1}\right) q^{i\nu\alpha}$.

$\{d_n\}$ and $\{k_n\}$ are approximations to the identity on $\hat{K}^*$ (i.e., they converge as measures to the measure with mass 1 concentrated at the identity character) and while $\{d_n\}$ is not a non-negative sequence, $\{k_n\}$ is non-negative.

Recall that Ψ_0 is characteristic function of $\mathfrak{D}^*$. It is easy to check that $\mathfrak{M}^{-1}d_n(x) = \sum_{\nu=-n}^{n} \Psi_0(\mathfrak{p}^\nu x)$, and $\mathfrak{M}^{-1}k_n(x) = \sum_{\nu=-n}^{n}\left(1-\frac{|\nu|}{n+1}\right)\Psi_0(\mathfrak{p}^\nu x)$.

From these observations we obtain:

$$P\int f = \lim_{n\to\infty}\int_K f(x)\mathfrak{M}^{-1}d_n(x)\,dx = \lim_{n\to\infty} < \mathfrak{M}(|x|f), d_n >$$

$$(C,1)\int f = \lim_{n\to\infty}\int_K f(x)\mathfrak{M}^{-1}k_n(x)\,dx = \lim_{n\to\infty} < \mathfrak{M}(|x|f), k_n > .$$

It is easy to see that if π is a continuous multiplicative character on K^* then $\pi = \pi^*|\cdot|^\beta$ where π^* is a (unitary) character on $\mathfrak{O}^*$ and $\beta \in C$. π is unitary iff $\mathrm{Re}(\beta) = 0$ and $\pi_1^*|\cdot|^\beta = \pi_1 = \pi_2 = \pi_2^*|\cdot|^\nu$ iff $\pi_1^* = \pi_2^*$ and $\beta - \nu = 2k\pi i/\ln q$ for some $k \in \mathbb{Z}$. In the sequel we restrict β to the strip, $-\pi/\ln q < \mathrm{Im}(\beta) \leq \pi/\ln q$. All the functions we consider may be extended to entire or meromorphic functions by periodicity.

5. Gamma, Beta and Bessel Functions. The Hankel transform

We define the gamma function $\Gamma(\pi) = \Gamma(\pi^*|\cdot|^\alpha) = \Gamma_{\pi^*}(\alpha)$ for all (not necessarily unitary) multiplicative characters π, except for the identity character $\pi \equiv 1$. If π^* is unramified we write $\pi^* = 1$.

Definition. (1) If π is ramified

$$\Gamma(\pi) = \Gamma_{\pi^*}(\alpha) = P\int \overline{\chi}(x)\pi(x)|x|^{-1}dx .$$

(2) If π is unramified:

(a) If $\text{Re}(\alpha) > 0$, $\Gamma(|\cdot|^{\alpha}) = \Gamma_1(\alpha) = P\int \overline{\chi}(x)|x|^{\alpha-1}dx$.

(b) If $\text{Re}(\alpha) = 0$, $\alpha \neq 0$, $\Gamma_1(\alpha) = (C,1)\int \overline{\chi}(x)|x|^{\alpha-1}dx$.

(c) If $\text{Re}(\alpha) < 0$, $\Gamma_1(\alpha)$ is the analytic continuation of Γ_1 into the left half plane as a meromorphic function.

In Theorem (5.5) we will show that these definitions make sense by explicitly evaluating the integrals in the definitions. The essential details of the computation are contained in the following three lemmas:

<u>Lemma (5.1)</u>. <u>Let</u> π <u>be ramified of degree</u> h, $h \geq 1$. <u>Then</u>

$$(5.2) \qquad \int_{|x|=1} \overline{\chi}(ux)\pi(x)\,dx : \begin{cases} = 0 & \text{if } |u| \neq q^h \\ \neq 0 & \text{if } |u| = q^h \end{cases} .$$

<u>Proof</u>. Let $f(x) = \pi(x)$ if $|x| = 1$, $f(x) = 0$ otherwise. Then $f \in \mathcal{S}$ and f is constant on cosets (in K^+) of $\mathfrak{P}^k$. Hence, by (3.2) $\hat{f}$ is supported on $\mathfrak{P}^{-k}$, so

$$\int_{|x|=1} \overline{\chi}(ux)\pi(x)\,dx = \int f(x)\overline{\chi}(ux)\,dx = \hat{f}(u) = 0 \quad \text{if} \quad |u| > q^h .$$

Let $g(x) = \overline{\chi}(ux)$ if $|x| = 1$, $g(x) = 0$ otherwise, $1 < q^k = |u|$. Then $g \in \mathcal{S}^*$ and g is constant on cosets (in K^*) of A_k. Hence by

(4.4) $\mathfrak{M}g$ is supported on A'_k, so if $h \geq k \geq 1$,

$$\int_{|x|=1} \bar{\chi}(ux)\pi(x)\,dx = \int g(x)\pi(x)\frac{dx}{|x|} = \mathfrak{M}g(\pi) = 0 .$$

If $|u| = q^k$, $k \leq 0$, then $u \in \mathfrak{O}$ so $\chi(ux) \equiv 1$ and

$$\int_{|x|=1} \bar{\chi}(ux)\pi(x)\,dx = \int_{|x|=1} \pi(x)\,dx = 0, \text{ since } \pi \text{ is ramified.}$$

Hence the integral in (5.2) is equal to zero if $|u| \neq q^h$. But the integral is $\hat{f}(u)$ and if $\hat{f}(u) = 0$ for some u, $|u| = q^h$ then $\hat{f}(u) \equiv 0$ and this implies $f \equiv 0$. But $f \not\equiv 0$ so $\hat{f}(u) \neq 0$, $|u| = q$

__Lemma (5.3).__

$$\int_{|x|=q^k} \bar{\chi}(x)\,dx = \begin{cases} q^k(1-q^{-1}), & k \leq 0 \\ -1, & k = 1 \\ 0, & k > 1 \end{cases} .$$

This is a formal restatement of an observation made in §1.

__Lemma (5.4).__ *If* $\deg(\pi) = h \geq 1$, *then*

$$\int_{A_k \sim A_{k+1}} \pi(x)\,dx = \begin{cases} 0, & 0 \leq k < h-1 \\ -q^{-h}, & k = h-1 \\ q^{-k}(1-q^{-1}), & k \geq h \end{cases} .$$

__Proof.__ This is the multiplicative analogue of (5.3). The proof is left as an easy exercise.

__Theorem (5.5)__. (i) __If__ $\deg(\pi) = h \geq 1$ __then__ $\Gamma(\pi) = \Gamma_{\pi^*}(\alpha) = c_{\pi^*}q^{h(\alpha-1/2)}$, __where__ $|c_{\pi^*}| = 1$, $c_{\pi^*}c_{\pi^{*-1}} = \pi^*(-1)$.

(ii) $\Gamma_1(\alpha) = (1-q^{\alpha-1})/(1-q^{-\alpha})$, $\alpha \neq 0$. $\Gamma_1(\alpha)$ __has a simple pole at__ $\alpha = 0$ __with residue__ $(1-q^{-1})\ell n\, q$. $1/\Gamma_1(\alpha)$ __has a simple pole at__ $\alpha = 1$ __with residue__ $-(1-q^{-1})\ell n\, q$. $\alpha = 0$ __is the only singularity of__ $\Gamma_1(\alpha)$; $\alpha = 1$ __is the only zero.__

(iii) __For all__ $\pi^* \in \hat{\mathfrak{O}}^*$, $\overline{\Gamma_{\pi^*}(\alpha)} = \pi^*(-1)\Gamma_{\pi^{*-1}}(\bar{\alpha})$; $\Gamma_{\pi^*}(\alpha)\Gamma_{\pi^{*-1}}(1-\alpha) = \pi^*(-1)$, __and so__ $\Gamma_{\pi^*}(\alpha)\overline{\Gamma_{\pi^*}(1-\bar{\alpha})} = 1$, __with the obvious interpretations at__ $\alpha = 0$ __or__ 1 __for__ $\pi^* = 1$.

__Proof__. If $\deg(\pi^*) = h \geq 1$, then by (5.1) we see that

$$\Gamma_{\pi^*}(\alpha) = \int_{|x|=q^h} \bar{\chi}(x)\pi(x)|x|^{-1}dx = q^{h(\alpha-\frac{1}{2})}\int_{|x|=q^h} \bar{\chi}(x)\pi^*(x)|x|^{\frac{1}{2}-1}dx$$

$$= \Gamma_{\pi^*}(1/2)q^{h(\alpha-\frac{1}{2})} = c_{\pi^*}q^{h(\alpha-\frac{1}{2})} \quad \text{with} \quad c_{\pi^*} = \Gamma_{\pi^*}(1/2).$$

Now suppose $\pi^* = 1$. Then,

$$\int_{q^{-n}\leq |x|\leq q^n} \bar{\chi}(x)|x|^{\alpha-1}dx = -q^{\alpha-1} + (1-q^{-1})\sum_{k=0}^{n} q^{-k\alpha}.$$

If $\mathrm{Re}(\alpha) > 0$ the limit is $-q^{\alpha-1} + (1-q^{-1})/(1-q^{-\alpha}) = (1-q^{\alpha-1})/(1-q^{-\alpha})$. If $\mathrm{Re}(\alpha) = 0$, $\alpha \neq 0$ the arithmetic averages are also $(1-q^{\alpha-1})/(1-q^{-\alpha})$. We continue this rational expression into the left half plane and an easy calculation (left as an exercise) establishes that the residues are as advertised.

<u>Proof of part (iii)</u>. For $\Gamma_1(\alpha)$ it is immediate. Now suppose $\deg(\pi^*) = h \geq 1$.

$$\overline{\Gamma_{\pi^*}(\alpha)} = \int_{|x|=q^h} \overline{\pi(x)}\ \overline{\chi}(-x) \frac{dx}{|x|} = \int_{|x|=q^h} \pi^{*-1}(x)|x|^{\overline{\alpha}}\ \overline{\chi}(-x) \frac{dx}{|x|}$$

$$= \int_{|x|=q^h} \pi^{*-1}(-x)|x|^{\overline{\alpha}}\ \overline{\chi}(x) \frac{dx}{|x|} = \pi^*(-1)\Gamma_{\pi^{*-1}}(\overline{\alpha}) \quad .$$

As in (5.1) we let $f(x) = \pi(x)\Psi_0$ and we obtain $\hat{f}(u) = \Gamma(\pi)\pi^{-1}(u)$ if $|u| = q^h$ and is zero otherwise. We also have that $(\hat{f})^{\vee}(x) = f(x)$ for all x. Thus,

$$\Gamma_{\pi^*}(\alpha)\Gamma_{\pi^{*-1}}(1-\alpha) = \Gamma(\pi) \int_{|u|=q^h} \pi^{*-1}(u)|u|^{1-\alpha}\ \overline{\chi}(u)|u|^{-1}du$$

$$= \int_{|u|=q^h} \Gamma(\pi)\pi^{-1}(u)\overline{\chi}(u)du = \int \hat{f}(u)\chi(-u) = f(-1)$$

$$= \pi(-1) = \pi^*(-1).$$

The rest of (iii) follows directly, and the properties of c_{π^*} in (i) follow from (iii) with $c_{\pi^*} = \Gamma_{\pi^*}(1/2)$.

We now show that the gamma function arises naturally as the normalizing factor of a certain distributuion.

Fix $\varphi \in \mathcal{S}$ and assume, without loss of generality, that φ is supported on $\mathfrak{P}^{-n}$ and is constant on cosets of $\mathfrak{P}^n$, $n \geq 1$. Let π be a ramified character. Then it is easily seen that

$$(P\pi|\cdot|^{-1},\varphi) = \int_{q^{-n}<|x|\leq q^n} \pi^*(x)\varphi(x)|x|^{\alpha-1}dx, \quad \pi = \pi^*|\cdot|^{\alpha}, \text{ is an}$$

analytic function of α. For π fixed it clearly defines a distribution. Now consider, for $\mathrm{Re}(\alpha) \geq 0$, $\alpha \neq 0$,

$$((c,1)|\cdot|^{\alpha-1},\varphi) = \varphi(0)(c,1)\int_{|x|\leq q^{-n}} |x|^{\alpha-1}dx + \int_{q^{-n}<|x|\leq q^n} \varphi(x)|x|^{\alpha-1}dx$$

$$= \varphi(0)q^{-n\alpha}(1-q^{-1})/(1-q^{-\alpha}) + \int_{q^{-n}<|x|\leq q^n} \varphi(x)|x|^{\alpha-1}dx .$$

As a function of α it extends to a meromorphic function with a simple pole at $\alpha = 0$, with residue $\varphi(0)(1-q^{-1})/\ln q$ at $\alpha = 0$. For $\alpha \neq 0$, it obviously defines a distribution.

We designate the distributions by $\pi|\cdot|^{-1}$.

Now notice that (using (5.5) (ii))

$(\Gamma_1(1-\alpha)|\cdot|^{\alpha-1},\varphi) = (1/\Gamma_1(\alpha)|\cdot|^{\alpha-1},\varphi)$ has a removable singularity at $\alpha = 0$, and its limiting value at $\alpha = 0$ is $\varphi(0) = (\delta,\varphi)$. It is thus natural to define (formally) $\Gamma_1(1)|\cdot|^{-1} = 1/\Gamma_1(0)|\cdot|^{-1} = \delta$ as elements of $\mathcal{S}'$.

Summarizing:

<u>Theorem (5.6)</u>. <u>If</u> $\pi \not\equiv 1$, <u>then</u> $\pi|\cdot|^{-1} \in \mathscr{S}'$.

<u>Theorem (5.7)</u>. <u>If</u> $\pi \not\equiv 1$ <u>or</u> $\pi(x) \neq |x|$ <u>then</u> $(\pi|\cdot|^{-1})\hat{} = \Gamma(\pi)\pi^{-1}$.

<u>Proof</u>. Write $\pi = \pi^*|\cdot|^{\alpha}$. We may assume that $0 < \mathrm{Re}(\alpha) < 1$ (using analytic continuation) and that φ is constant on cosets of $\mathfrak{P}^n$ and is supported on $\mathfrak{P}^{-n}$ for some $n \geq 1$, and so by (3.2) $\hat{\varphi}$ has the same properties. We need to show that $(\Gamma(\pi)\pi^{-1},\varphi) = (\pi|\cdot|^{-1},\hat{\varphi})$.

Suppose first that π is unramified, $\pi(x) = |x|^{\alpha}$.

$$(\Gamma(\pi)\pi^{-1},\varphi) = \varphi(0)q^{-n(1-\alpha)}\frac{1-q^{-1}}{\cancel{1-q^{\alpha-1}}}\cdot\frac{\cancel{1-q^{\alpha-1}}}{1-q^{-\alpha}} + \Gamma(\pi)\int_{q^{-n}<|x|\leq q^n}\varphi(x)\pi^{-1}(x)d$$

$$(\pi|\cdot|^{-1},\hat{\varphi}) = \int_{|u|\leq q^n}\pi(u)|u|^{-1}\hat{\varphi}(u)du = \int_{|u|\leq q^n}\pi(u)|u|^{-1}\int_{|x|\leq q^n}\varphi(x)\bar{\chi}(ux)dx\ d$$

$$= \int_{|x|\leq q^{-n}}\varphi(0)\int_{|u|\leq q^n}\pi(u)|u|^{-1}dudx$$

$$+ \int_{q^{-n}<|x|\leq q^n}\varphi(x)\int_{|u|\leq q^n}\pi(u)|u|^{-1}\chi(-ux)du\ dx\ .$$

$$\int_{|u|\leq q^n}\pi(u)|u|^{-1}du = \int_{|u|\leq q^n}|u|^{\alpha-1}du = q^{n\alpha}(1-q^{-1})/(1-q^{-\alpha})$$

and we get that the first integral is equal to $\varphi(0)q^{-n(1-\alpha)}(1-q^{-1})/(1-q^{-\alpha}$

In the second integral let $y = ux$. We have it is equal to

$$\int_{q^{-n}<|x|\le q^n} \varphi(x)\pi^{-1}(x)\Big[\int_{|y|\le q^n|x|} \pi(y)\bar{\chi}(y)\frac{dy}{|y|}\Big]dx$$

$$= \Gamma(\pi) \int_{q^{-n}<|x|\le q^n} \varphi(x)\pi^{-1}(x)\,dx,$$

since $q^n|x| \ge q$, and the result is established for π unramified.

Suppose now that $\deg(\pi) = h \ge 1$. From (4.3) we see that φ is constant on cosets of A_{2n} throughout K^* and on cosets of A_{n-k} in $\mathfrak{P}^k$ for $k \le n$.

If $h > 2n$, $-n < k \le n$, then

$$\int_{|x|=q^k} \varphi(x)\pi^{-1}(x)\,dx = \sum \varphi(\beta_s) \int_{\beta_s A_{2n}} \pi^{-1}(x)\,dx = 0,$$

where the sum is over the finite collection of cosets of A_{2n} in $\{|x| = q^k\}$.

If $h \le 2n$, $-n < k < -n + h$

$$\int_{|x|=q^k} \varphi(x)\pi^{-1}(x)\,dx = \sum \varphi(\beta_s) \int_{\beta_s A_{n+k}} \pi^{-1}(x)\,dx = 0 .$$

Thus,

$$(\Gamma(\pi)\pi^{-1},\varphi) = \Gamma(\pi) \int_{q^{-n}<|x|\le q^n} \varphi(x)\pi^{-1}(x)\,dx = \begin{cases} 0 & , h > 2n \\ \Gamma(\pi) \displaystyle\int_{q^{-n+h}\le|x|\le q^n} \varphi(x)\pi^{-1}(x)\,dx , & 1 \le h \le 2n . \end{cases}$$

On the other hand,

$$(\pi|\cdot|^{-1},\hat{\varphi}) = \int\limits_{q^{-n}<|u|\le q^n} \pi(u)|u|^{-1}\hat{\varphi}(u)\,du$$

$$= \int\limits_{q^{-n}<|u|\le q^n} \pi(u)|u|^{-1} \int\limits_{|x|\le q^n} \varphi(x)\chi(-ux)\,dx\;du$$

$$= \int\limits_{q^{-n}<|u|\le q^n} \pi(u)|u|^{-1} \int\limits_{q^{-n}<|x|\le q^n} \varphi(x)\chi(-ux)\,dx\;du$$

$$= \int\limits_{q^{-n}<|x|\le q^n} \varphi(x) \int\limits_{q^{-n}<|u|\le q^n} \pi(u)|u|^{-1}\chi(-ux)\,du\;dx$$

$$= \int\limits_{q^{-n}<|x|\le q^n} \varphi(x)\pi^{-1}(x) \int\limits_{q^{-n}|x|<|y|\le q^n|x|} \bar{\chi}(y)\pi(y)\,\frac{dy}{|y|}\,dx\ .$$

Note that $q^n|x| \le q^{2n}$, so that if $h > 2n$

$$\int\limits_{q^{-n}|x|<|y|\le q^n|x|} \bar{\chi}(y)\pi(y)\,\frac{dy}{|y|} = 0\ .$$

On the other hand suppose $1 \le h \le 2n$. If $q^{-n+h} \le |x| \le q^n$ then $|y| = q^h$ is included in the range of integration so

$$\int\limits_{q^{-n}|x|<|y|\le q^n|x|} \bar{\chi}(y)\pi(y)\,\frac{dy}{|y|} = \Gamma(\pi)\ .$$

For $q^{-n} < |x| < q^{-n+h}$, the range of integration does not contain $|y| = q^h$ so $\displaystyle\int\limits_{q^{-n}|x|<|y|\le q^n|x|} \bar{\chi}(y)\pi(y)\,\frac{dy}{|y|} = 0\ .$

It now follows that $(\pi|\cdot|^{-1},\hat{\varphi}) = (\Gamma(\pi)\pi^{-1},\varphi)$.

We now examine another approach to the proof of (5.7).

<u>Definition</u>. $f \in \mathscr{S}'$ is said to be <u>homogeneous of degree π</u> (π a multiplicative character) if for all $t \in K^*$, $f_t = \pi(t)f$; where for functions $f_t(x) = f(tx)$ and for distributions $(f_t,\varphi) = (f,|t|^{-1}\varphi_{t^{-1}})$, $\varphi \in \mathscr{S}$.

<u>Examples</u>. If π is a multiplicative character and $\pi \neq |\cdot|^{-1}$, then $\pi \in \mathscr{S}'$ and π is homogeneous of degree π (see the definition π as an element of $\mathscr{S}'$). Furthermore δ is homogeneous of degree $|\cdot|^{-1}$. To see this, $(\delta_t,\varphi) = (\delta,|t|^{-1}\varphi_{t^{-1}}) = |t|^{-1}\varphi(0) = (|t|^{-1}\delta,\varphi)$.

<u>Lemma (5.8)</u>. <u>If</u> $f \in \mathscr{S}'$ <u>is supported at</u> 0 <u>then</u> $f = c\delta$ <u>for some</u> $c \in C$.

<u>Proof</u>. Since φ is constant on some $\mathfrak{P}^\ell$ and $(f,\varphi) = (f,\Phi_k\varphi)$ for all k it follows that (f,φ) is a function of $\varphi(0)$. But (f,φ) is linear so for some $c \in C$, $(f,\varphi) = c\varphi(0) = (c\delta,\varphi)$, and $f = c\delta$.

<u>Lemma (5.9)</u>. <u>If</u> $f \in \mathscr{S}'$ <u>is homogeneous of degree</u> π <u>then</u> $\hat{f}$ <u>is homogeneous of degree</u> $\pi^{-1}|\cdot|^{-1}$.

<u>Proof</u>. First note that if $f \in L^1$ then $(|t|^{-1}f_{t^{-1}})^{\wedge} = (\hat{f})_t$, by a change of variables in the defining integral. Thus, if $\varphi \in \mathscr{S}$,

$$((\hat{f})_t,\varphi) = (\hat{f},|t|^{-1}\varphi_{t^{-1}}) = (f,(|t|^{-1}\varphi_{t^{-1}})\hat{}) = (f,(\hat{\varphi})_t)$$

$$= |t|^{-1}(f,|t|(\hat{\varphi})_t) = |t|^{-1}(f_{t^{-1}},\hat{\varphi}) = |t|^{-1}\pi^{-1}(t)(f,\hat{\varphi})$$

$$= (|t|^{-1}\pi^{-1}(t)\hat{f},\varphi).$$

Lemma (5.10). *If* $f \in \mathcal{S}'$ *is homogeneous of degree* π *and* $\pi \neq |\cdot|^{-1}$ *then* $f = c\pi$ *for some* $c \in C$. *If* $f \in \mathcal{S}'$ *is homogeneous of degree* $|\cdot|^{-1}$ *then* $f = c\delta$ *for some* $c \in C$.

Proof. We assume that $f \in \mathcal{S}'$, f is homogeneous of degree π, $\pi \neq |\circ|^{-1}$. Recall that Ψ_n is the characteristic function of A_n, and so $(\Psi_n)_{\beta^{-1}}$ is the characteristic function of βA_n. Now note that $(f,(\Psi_n)_{\beta^{-1}}) = |\beta|(f_\beta,\Psi_n) = \pi(\beta)|\beta|(f,\Psi_n)$.

Suppose $\varphi \in \mathcal{S}$, $\varphi(0) = 0$. Then φ can be written as a finite linear combination of functions of the form $(\Psi_n)_{\beta^{-1}}$, with $n \geq 1$. For such a function we have $(\pi,(\Psi_n)_{\beta^{-1}}) = |\beta|\pi(\beta)q^{-n}$, provided we choose $n \geq \deg(\pi)$. If $\deg(\pi) = h \geq 1$, $(f,\Psi_n) = q^{-n}q^h(f,\Psi_h)$. If $\deg(\pi) = 0$, $n \geq 1$, $(f,\Psi_n) = (1-q^{-1})^{-1}(f,\Psi_0)q^{-n}$, and so $(f,(\Psi_n)_{\beta^{-1}}) = c|\beta|\pi(\beta)q^{-n}$ for some $c \in C$. Thus $(f-c\pi,\varphi) = 0$ for all $\varphi \in \mathcal{S}$ such that $\varphi(0) = 0$. But this means that $f-c\pi = c_1\delta$ for $c_1 \in C$. If $c_1 \neq 0$ the homogeneity is wrong so $c_1 = 0$, and $f = c\pi$.

If f is homogeneous of degree $|\cdot|^{-1}$ then (5.9) shows that $\hat{f}$ is homogeneous of degree $|\cdot|^{0}$. By the first part of the proof we get that $\hat{f}$ is a constant function and so $f = c\delta$ for some $c \in \mathbb{C}$.

<u>Another proof of (5.7).</u> Consider $\pi|\cdot|^{-1}$ with $\pi \not\equiv 1$ and $\pi \neq |\cdot|$. Then $(\pi|\cdot|^{-1})^{\wedge}$ is homogeneous of degree π^{-1} by (5.9). Thus, by (5.10) $(\pi|\cdot|^{-1})^{\wedge} = A_{\pi}\pi^{-1}$. From the definition of the Fourier transform $(\pi|\cdot|^{-1},\hat{\varphi}) = A_{\pi}(\pi^{-1},\varphi)$ for all $\varphi \in \mathscr{S}$.

Note that if $\varphi = q^{k}\tau_{1}\Phi_{k}$ then $\hat{\varphi} = \bar{\chi}\,\Phi_{-k}$.

Suppose $\deg(\pi) = h \geq 1$. Then

$$\Gamma(\pi) = (\pi|\cdot|^{-1},\bar{\chi}\Phi_{-h}) = A_{\pi}(\pi^{-1},q^{-k}\tau_{1}\Phi_{h}) = A_{\pi}.$$

If $\deg(\pi) = 0$ and $\pi = |\cdot|^{\alpha}$, $0 < \mathrm{Re}(\alpha) < 1$, we get $\Gamma(\pi) = (|\cdot|^{\alpha-1},\bar{\chi}\Phi_{-1}) = A_{\pi}(|\cdot|^{-\alpha},q^{-1}\tau_{1}\Phi_{1}) = A_{\pi}$. The result extends to $\alpha \neq 0$ by analytic continuation, and the proof is complete.

<u>Definition.</u> Let $\pi = \pi^{*}|\cdot|^{\alpha}$, $\lambda = \lambda^{*}|\cdot|^{\beta}$ be multiplicative characters. The <u>beta function</u>, $B(\pi,\lambda)$ is defined as

$$B(\pi,\lambda) = \frac{\Gamma(\pi)\Gamma(\lambda)}{\Gamma(\pi\lambda)} = \frac{\Gamma_{\pi^{*}}(\alpha)\Gamma_{\lambda^{*}}(\beta)}{\Gamma_{\pi^{*}\lambda^{*}}(\alpha+\beta)}.$$

$B(\pi,\lambda)$ can be considered as a meromorphic function of the two complex variables α,β for π^{*},λ^{*} fixed. In a number of cases $B(\pi,\lambda)$ is constant as a function of α,β or both. Note that

$\deg(\pi^*\lambda^*) \leq \max[\deg(\pi^*), \deg(\lambda^*)]$ and if $\deg(\pi^*) \neq \deg(\lambda^*)$ then $\deg(\pi^*\lambda^*) = \max[\deg(\pi^*), \deg(\lambda^*)]$. Thus, if π is unramified, λ is ramified of degree $h \geq 1$, then $\deg(\pi\lambda) = h$ and $B(\pi,\lambda) = \Gamma_1(\alpha)q^{-h\alpha}$. If $\deg(\pi) = h_1 \geq 1$, $\deg(\lambda) = h_2 \geq 1$, $h_1 \neq h_2$, then $B(\pi,\lambda) = (c_{\pi^*}c_{\lambda^*}/c_{\pi^*\lambda^*})q^{(h_1-h_2)\alpha-h_1/2}$. If $\deg(\pi) = \deg(\lambda) = \deg(\pi\lambda)=h \geq 1$ then $B(\pi,\lambda) = (c_{\pi^*}c_{\lambda^*}/c_{\pi^*\lambda^*})q^{-h/2}$. Clearly $B(\pi,\lambda) = B(\lambda,\pi)$.

Our aim is to obtain an integral representation of the beta function under suitable conditions.

Let us assume:

$$(5.11) \qquad 0 < \mathrm{Re}(\alpha), \mathrm{Re}(\beta), \mathrm{Re}(\alpha+\beta) < 1, \quad \pi = \pi^*|\cdot|^{\alpha}, \quad \lambda = \lambda^*|\cdot|^{\beta}.$$

If $u \in K^*$ the integral

$$(5.12) \qquad k(u) = (\pi|\cdot|^{-1} * \lambda|\cdot|^{-1})(u) = \int \pi(x)|x|^{-1}\lambda(u-x)|u-x|^{-1}dx$$

converges absolutely. If we define

$$(5.13) \qquad b(\pi,\lambda) = \int \pi(x)|x|^{-1}\lambda(1-x)|1-x|^{-1}dx = k(1) \quad ,$$

and change variables in (5.12) we see that $k(u) = (\pi\lambda)(u)|u|^{-1}b(\pi,\lambda)$. By (5.7) we have that $\hat{k} = b(\pi,\lambda)\Gamma(\pi\lambda)(\pi\lambda)^{-1}$. We wish to show that $b(\pi,\lambda) = B(\pi,\lambda)$ which will follow from the relation $\hat{k} = \Gamma(\pi)\Gamma(\lambda)(\pi\lambda)^{-1}$. That is the substance of the following lemma.

<u>Lemma (5.14)</u>. <u>If</u> π, λ <u>satisfy</u> (5.11) <u>then</u> $(\pi|\cdot|^{-1} * \lambda|\cdot|^{-1})^{\wedge}$
$= (\pi|\cdot|^{-1})^{\wedge}(\lambda|\cdot|^{-1})^{\wedge} = \Gamma(\pi)\Gamma(\lambda)(\pi\lambda)^{-1}$.

<u>Proof</u>. For g a locally integrable function (on K^*) let $[g]_n(x) = g(x)$ if $q^{-n} \leq |x| \leq q^n$, and zero otherwise.

If $n \geq 1$ we see that $[\pi|\cdot|^{-1}]_n * \lambda|\cdot|^{-1} \in \mathscr{S}'$ since it is bounded by a constant (independent of n) multiple of $|u|^{\mathrm{Re}(\alpha+\beta)-1}$, which is locally integrable.

Fix $\varphi \in \mathscr{S}$ such that $\varphi(0) = 0$. Then for some k, $k \geq 1$ φ is supported on $\{q^{-k} < |x| \leq q^k\}$ and is constant on cosets of $\mathfrak{P}^k$. Note that $\hat{\varphi}$ is supported on $\mathfrak{P}^{-k}$. Thus,

(5.15) $$(([\pi|\cdot|^{-1}]_n * \lambda|\cdot|^{-1})^{\wedge}, \varphi) = ([\pi|\cdot|^{-1}]_n * \lambda|\cdot|^{-1}, \hat{\varphi})$$

$$= \int_{|u| \leq q^k} \hat{\varphi}(u) \int_{q^{-n} \leq |x| \leq q^n} \pi(x)|x|^{-1} \lambda(u-x)|u-x|^{-1} dx\, du$$

$$= \int_{q^{-n} \leq |x| \leq q^n} \pi(x)|x|^{-1} \int_{|x| \leq q^k} \lambda(u-x)|u-x|^{-1}\hat{\varphi}(u)\, du\, dx .$$

$$\int_{|u| \leq q^k} \lambda(u-x)|u-x|^{-1} \hat{\varphi}(u)\, du = (\tau_x(\lambda|\cdot|^{-1}), \hat{\varphi})$$

$$= ((\tau_x(\lambda|\cdot|^{-1}))^{\wedge}, \varphi) = (\Gamma(\lambda)\bar{\chi}_x \lambda^{-1}, \varphi) \quad \text{(by (5.7))}$$

$$= \int_{q^{-k} < |u| \leq q^k} \Gamma(\lambda)\bar{\chi}_x(u)\lambda^{-1}(u)\varphi(u)\, du .$$

By Fubini's theorem we have,

$$(([\pi|\cdot|^{-1}]_n * \lambda|\cdot|^{-1})^{\wedge},\varphi) =$$

$$= \int_{q^{-k}<|u|\leq q^k} \Gamma(\lambda)\lambda^{-1}(u)\varphi(u) \int_{q^{-n}\leq|x|\leq q^n} \pi(x)|x|^{-1}\bar{\chi}(ux)\,dx\,du$$

$$= \int_{q^{-k}<|u|\leq q^k} \Gamma(\lambda)(\pi\lambda)^{-1}(u)\varphi(u) \int_{q^{-n}|u|^{-1}\leq|x|\leq q^n|u|^{-1}} \pi(x)|x|^{-1}\bar{\chi}(x)\,dx\,d$$

For n large enough the inner integral is $\Gamma(\pi)$, so

$$\lim_{n\to\infty} ((\pi|\cdot|^{-1}]_n * \lambda|\cdot|^{-1})^{\wedge},\varphi) = (\Gamma(\pi)\Gamma(\lambda)(\pi\lambda)^{-1},\varphi) \quad \text{provided} \quad \varphi(0) = 0.$$

From (5.8) we have that $\lim_{n\to\infty} ([\pi|\cdot|^{-1}]_n * \lambda|\cdot|^{-1})^{\wedge} = \Gamma(\pi)\Gamma(\lambda)(\pi\lambda)^{-1} + c_1\delta$ (
$c_1 \in C$.

On the other hand if $\varphi \in \mathscr{S}$, $\varphi(0) = 0$ then

$$([\pi|\cdot|^{-1}]_n * \lambda|\cdot|^{-1},\varphi) = \int_{q^{-k}<|u|\leq q^k} \varphi(u) \int_{q^{-n}\leq|x|\leq q^n} \pi(x)|x|^{-1}\lambda(u-x)|u-x|^{-1}dx$$

$$= \int_{q^{-k}<|u|\leq q^k} \varphi(u)(\pi\lambda)(u)|u|^{-1} \int_{q^{-n}|u|^{-1}\leq|x|\leq q^n|u|^{-1}} \pi(x)|x|^{-1}\lambda(1-x)|1-x|^{-1}dx$$

$\to b(\pi,\lambda)(\pi\lambda|\cdot|^{-1},\varphi)$ as $n \to \infty$, since $\varphi(u)(\pi\lambda)(u)|u|^{-1} \in L^1$ and the inner integral converges uniformly to

$b(\pi,\lambda)$ on the support of that function. Thus,

$$\lim_{n\to\infty} ([\pi|\cdot|^{-1}]_n * \lambda|\cdot|^{-1}) = b(\pi,\lambda)\pi\lambda|\cdot|^{-1} + c_2\delta(\mathscr{S}'), \; c_2 \in C$$ and hence

$$\lim_{n\to\infty} ([\pi|\cdot|^{-1}]_n * \lambda|\cdot|^{-1})^\wedge = b(\pi,\lambda)\Gamma(\pi,\lambda)(\pi\lambda)^{-1} + c_2 .$$ Thus,

$\Gamma(\pi)\Gamma(\lambda)(\pi\lambda)^{-1} + c_1\delta = b(\pi,\lambda)\Gamma(\pi\lambda)(\pi\lambda)^{-1} + c_2$ in $\mathscr{S}'$. By homogeneity we see that $c_1 = c_2 = 0$ and the result follows.

The conclusion of (5.14) is summarized:

<u>Theorem (5.16)</u>. <u>If (5.11) is satisfied, then</u>

$$B(\pi,\lambda) = \int \pi(x)|x|^{-1} \lambda(1-x)|1-x|^{-1}dx .$$

<u>Furthermore</u>, $[\pi|\cdot|^{-1}]_n * \lambda|\cdot|^{-1} \to B(\pi,\lambda)\pi\lambda|\cdot|^{-1}$ <u>in</u> $\mathscr{S}'$.

<u>Remark.</u> A variety of extensions of (5.16) are possible using P-integrals and (C,1)-integrals. We shall avoid them.

<u>Corollary (5.17)</u>. <u>If</u> $0 < \mathrm{Re}(\alpha), \mathrm{Re}(\beta), \mathrm{Re}(\alpha+\beta) < 1$ <u>then</u>

$$\left(\frac{1}{\Gamma_1(\alpha)}|x|^{\alpha-1}\right) * \left(\frac{1}{\Gamma_1(\beta)}|x|^{\beta-1}\right) = \frac{1}{\Gamma_1(\alpha+\beta)}|x|^{\alpha+\beta-1} .$$

<u>Notation.</u> The delta function, δ, always refers to a measure of mass 1 concentrated at the origin. Thus, in $K(K^+)$ it is the δ-function

described earlier. In the context of $\hat{K}^*$ it is concentrated at the identity character.

We now show that the gamma function, for unitary characters π, may be considered (up to a δ-function) as the Mellin transform of $\overline{X} \in \mathcal{S}^{*\prime}$ evaluated at π.

<u>Definition</u>. $\Gamma^{\#} = \mathfrak{M}\,\overline{X} \in \hat{\mathcal{S}}^{*\prime}$.

The structure of $\hat{K}^*$ is that of a countable, discrete collection of circles: $\{\mathbb{T}_{\pi^*}\}_{\pi^* \in \hat{\mathfrak{O}}^*}$. Hence $\Gamma^{\#}$ may be viewed as a collection $\Gamma^{\#} \sim \{\Gamma^{\#}_{\pi^*}\}$ where $\Gamma^{\#}_{\pi^*}$ is a distribution on $\mathbb{T}_{\pi^*}$, a copy of the circle.

Observe that $P\Gamma_1(i\alpha)$ and $\Gamma_{\pi^*}(i\alpha)$ $(\deg(\pi^*) > 0)$, induce distributions on the circle. If π^* is ramified, $\deg(\pi^*) = h \geq 1$, then $\Gamma_{\pi^*}(i\alpha) = c_{\pi^*}\, q^{h/2}\, q^{i\alpha/2}$, which is a distribution on $\mathbb{T}_{\pi^*}$ since it is bounded.

$$\Gamma_1(i\alpha) = \frac{1-q^{i\alpha-1}}{1-q^{-i\alpha}} + q^{i\alpha-1} - q^{i\alpha-1} = \frac{1-q^{-1}}{2i}\,\frac{q^{i\alpha/2}}{\sin(\alpha\,\ln q/2)} - q^{i\alpha-1}$$

Clearly $P\Gamma_1(i\alpha)$ is a distribution on $\mathbb{T}_1$ (not unrelated to the conjugate kernel).

Explicitly, these distributions are given by:

$$<\Gamma_{\pi^*},\varphi> = \frac{\ln q}{(1-q^{-1})}\frac{1}{2\pi}\int_{-\pi/\ln q}^{\pi/\ln q}\Gamma_{\pi^*}(i\alpha)\varphi(\alpha)\,d\alpha\,,\ \deg(\pi^*) > 0$$

$$<P\Gamma_1,\varphi> = \lim_{\epsilon\to 0}\frac{\ln q}{(1-q^{-1})}\frac{1}{2\pi}\int_{\epsilon\le|\alpha|\le\pi/\ln q}\Gamma_1(i\alpha)\varphi(\alpha)\,d\alpha\,,$$

for φ a "trigonometric polynomial" on $\mathbb{T}$. That is, $\varphi(\alpha) = \sum_{-n}^{n} a_\nu q^{i\nu\alpha}$.

We write $\Gamma \approx \{P\Gamma_1\,;\,\Gamma_{\pi^*}\,,\ \deg(\pi^*) > 0\} \in \hat{\mathcal{S}}^{*\prime}$.

<u>Theorem (5.19)</u>. (i) $\Gamma_1^{\#} = P\Gamma_1 + (\pi(1-q^{-1})/\ln q)\delta$.

(ii) $\Gamma_{\pi^*}^{\#} = \Gamma_{\pi^*}\,,\ \deg(\pi^*) > 0$.

<u>Proof</u>. Let $\varphi \in \hat{\mathcal{S}}^*$. Then $\mathfrak{M}^{-1}\varphi \in \mathcal{S}^*$ and is compactly supported on K^*. Thus, $<\Gamma^{\#},\varphi> = <\mathfrak{M}\bar{\chi},\varphi> = \lim_{n\to\infty} <\mathfrak{M}[\bar{\chi}]_n,\varphi>$. $[\bar{\chi}]_n \in \mathcal{S}^*$ and

$$\mathfrak{M}[\bar{\chi}]_n = \int_{q^{-n}\le|x|\le q^n}\bar{\chi}(x)\pi(x)|x|^{-1}dx\ .$$

If $\deg(\pi^*) \ge 1$ we see that $\mathfrak{M}[\bar{\chi}]_n(\pi^*|\cdot|^{i\alpha}) = \Gamma_{\pi^*}(i\alpha)$ if $n \ge \deg(\pi^*)$ and so $\Gamma_{\pi^*}^{\#} = \Gamma_{\pi^*}$ if π^* is ramified.

If $\pi^* = 1$ then,

$$\mathfrak{M}[\bar{\chi}]_n(|\cdot|^{i\alpha}) = \int_{q^{-n}\le|x|\le q^n} \bar{\chi}(x)|x|^{i\alpha-1}dx = -q^{i\alpha-1} + (1-q^{-1})\sum_{k=0}^{n} q^{ik\alpha}$$

$$= -q^{i\alpha-1} + \tfrac{1}{2}(1-q^{-1}) + (1-q^{-1})[D_n(\alpha\, \ln q) - i\,\tilde{D}_n(\alpha\, \ln q)$$

where D_n and $\tilde{D}_n$ are the Dirichlet and conjugate Dirichlet kernels: $D_n(x) = \frac{1}{2} + \sum_1^n \cos kx$, $\tilde{D}_n(x) = \sum_1^n \sin kx$. Well known facts from the theory of Fourier series show that as distributions on the circle $D_n(\alpha\, \ln q) \to (\pi/\ln q)\delta$ and $\tilde{D}_n(\alpha\, \ln q) \to P[\frac{1}{2}\cot(\alpha\, \ln q/2)]$. A little arithmetic shows that

$$-q^{i\alpha-1} + \tfrac{1}{2}(1-q^{-1})[1-i\cot(\alpha\, \ln q/2)] = \Gamma_1(i\alpha) \quad .$$

Thus, $\mathfrak{M}[\bar{\chi}]_n(|\cdot|^{i\alpha}) \to P\Gamma_1(i\alpha) + (\pi(1-q^{-1})/\ln q)\delta\ (\mathcal{S}^{*\prime})$ so $\Gamma_1^{\#} = P\Gamma_1 + (\pi(1-q^{-1})/\ln q)\delta$.

<u>Definition</u>. For $\pi \in \hat{K}^*$; $u,v \in K^*$, the <u>Bessel function (of order π)</u> denoted $J_\pi(u,v)$, is the value of the principal value integral

$$(5.20) \qquad P\int \bar{\chi}\,(ux + v/x)\pi(x)|x|^{-1}dx \quad .$$

In Theorem (5.25) below we establish that $J_\pi(u,v)$ exists for all $\pi \in \hat{K}^*$ and $u,v \in K^*$. If we use that fact several properties of the Bessel function can be obtained by changing variables in (5.20).

Lemma (5.21). (i) $J_\pi(u,v) = J_{\pi^{-1}}(v,u)$.

(ii) $\pi(u)J_\pi(u,v) = \pi(v)J_\pi(v,u)$.

(iii) $J_\pi(u,v) = \overline{J_{\pi^{-1}}(-u,-v)} = \pi(-1)\overline{J_{\pi^{-1}}(u,v)}$.

(iv) If $\pi(-1) = 1$, $J_\pi(u,u)$ is real-valued.

If $\pi(-1) = -1$, $J_\pi(u,u)$ is pure imaginary-valued.

For k, a positive integer, $\pi \in \hat{K}^*$, $v \in K^*$ we set

$$F_\pi(k,v) = \int_{|x|=q^k} \bar{\chi}(x)\bar{\chi}(v/x)\pi(x)|x|^{-1}dx \quad . \tag{5.22}$$

Lemma (5.23). Suppose $|v| = q^m$, $1 \leq k < m$

(i) If π is unramified, $F_\pi(k,v) \neq 0$ iff m is even and $k = m/2$.

(ii) If π is ramified of degree $h \geq 1$, $F_\pi(k,v) \neq 0$ iff one of the following holds: (a) m is even, $m \geq 2h$ and $k = m/2$,

(b) m is even, $h < m < 2h$ and $(k = m/2,\ k = h$ or $k = m-h)$

(c) m is odd, $h < m < 2h$ and $(k = h$ or $k = m-h)$.

Proof. Set $f_1(x) = \begin{cases} \pi(x)\bar{\chi}(x), & |x|=q^k \\ 0, & \text{otherwise} \end{cases}$; $f_2(x) = \begin{cases} \bar{\chi}(x), & |x| = q^{m-k} \\ 0, & \text{otherwise} \end{cases}$.

Then $F_\pi(k,v) = (f_1 * f_2)(v)$ where convolution is taken with respect to the multiplicative structure (K^*, dx^*). It follows that

$\mathfrak{M}(f_1 * f_2) = (\mathfrak{M}f_1)\cdot(\mathfrak{M}f_2)$. It follows that $F_\pi(k,v) \neq 0$ iff there is a $\pi' \in \hat{K}^*$ such that $(\mathfrak{M}f_1)(\pi')(\mathfrak{M}f_2)(\pi') \neq 0$. From our calculations for the gamma function (namely (5.1), (5.3) and (5.5)) we see that $\mathfrak{M}f_1(\pi') \neq 0$ iff either $\pi\pi'$ if unramified and $k = 1$ or $\pi\pi'$ is ramified of degree k. Similarly $\mathfrak{M}f_2(\pi') \neq 0$ iff either π' is unramified and $k = m-1$ or π' is ramified of degree $m-k$.

A straight forward check of the possibilities for π' gives our result. We provide the details for π unramified. Then there are two possibilities. Either $\pi\pi'$ and π' are both ramified of the same order, or both are unramified. If $\mathfrak{M}f_1(\pi')\mathfrak{M}f_2(\pi') \neq 0$ and π' is unramified then $k = 1$ and $k = m-1$ and $m = 2$, even. If π' is ramified of degree h then $k = h$ and $h = m-k$ so $m = 2h$, even. Thus, $F_\pi(k,v) = 0$ if m is not even, or $k \neq m/2$. Suppose now that $m > 1$ m is even, $k = m/2$. For $m = 2$ let $\pi' = 1$. If $m \geq 4$ choose any π' such that $\deg(\pi') = k$. This completes the proof for π unramified since such π' exist. The details (which are rather messy) are left as an exercise.

Lemma (5.24). $|F_\pi(k,v)| \leq (1-q^{-1})$.

Proof. $$|F_\pi(k,v)| \leq \int_{|x|=q^k} 1\,|x|^{-1}dx = 1-q^{-1}.$$

<u>Theorem (5.25)</u>. Let $\pi \in \hat{K}^*$, $u,v \in K^*$

(i) (a) <u>If</u> π <u>is unramified</u>, $\pi \not\equiv 1$

$$J_\pi(u,v) = \begin{cases} \pi(v)\Gamma(\pi^{-1}) + \pi^{-1}(u)\Gamma(\pi), & |uv| \le q \\ \pi^{-1}(u)F_\pi(m/2,uv), & |uv| = q^m,\ m > 1,\ m \text{ even} \\ 0 & |uv| = q^m,\ m > 1,\ m \text{ odd} \end{cases}$$

(b) <u>If</u> $\pi \equiv 1$

$$J_1(u,v) = \begin{cases} (1-q^{-1})[\log_q \frac{1}{|uv|} + 1] - 2/q, & |uv| \le q \\ F_\pi(m/2,uv), & |uv| = q^m,\ m > 1,\ m \text{ even} \\ 0 & |uv| = q^m,\ m > 1,\ m \text{ odd} \end{cases}$$

(ii) <u>If</u> π <u>is ramified</u>, $\deg(\pi) = h \ge 1$ <u>then</u>

$$J_\pi(u,v) = \begin{cases} \pi(v)\Gamma(\pi^{-1}) + \pi^{-1}(u)\Gamma(\pi); & |uv| \le q^h \\ \pi^{-1}(u)F_\pi(m/2,uv); & |uv| = q^m,\ m \ge 2h,\ m \text{ even} \\ 0 ; & |uv| = q^m,\ m > 2h,\ m \text{ odd} \\ \pi^{-1}(u)[F_\pi(h,uv) + F_\pi(m-h,uv) + F_\pi(m/2,uv)]; & \\ \qquad |uv| = q^m,\ h < m < 2h,\ m \text{ even} & \\ \pi^{-1}(u)[F_\pi(h,uv) + F_\pi(m-h,uv)]; & \\ \qquad |uv| = q^m,\ h < m < 2h,\ m \text{ odd.} & \end{cases}$$

<u>Proof</u>. Suppose $|uv| \le q$, $|uv| = q^m$, $m \le 1$.

$$J_\pi(u,v) = \pi^{-1}(u)P\int_{|x|\leq 1} \bar{\chi}(uv/x)\pi(x)|x|^{-1}dx + \pi^{-1}(u)P\int_{|x|>1} \bar{\chi}(x)\pi(x)|x|^{-1}dx$$

$$= \pi(v)P\int_{|x|\geq|uv|} \bar{\chi}(x)\pi^{-1}(x)|x|^{-1}dx + \pi^{-1}(u)P\int_{|x|>1}\bar{\chi}(x)\pi(x)|x|^{-1}dx$$

If π is ramified we see that $J_\pi(u,v) = \pi(v)\Gamma(\pi^{-1}) + \pi^{-1}(u)\Gamma(\pi)$.

If π is unramified set $\pi(x) = |x|^\alpha$ and recall that $\mathrm{Re}(\alpha) = 0$. Then

$$J_\pi(u,v) = |v|^\alpha \int_{q^m\leq|x|\leq q} \bar{\chi}(x)|x|^{-\alpha-1}dx + |u|^{-\alpha}\int_{|x|=q}\bar{\chi}(x)|x|^{\alpha-1}dx$$

$$= |v|^\alpha\{(1-q^{-1})\sum_{k=m}^{0} q^{-k\alpha} - q^{-\alpha-1}\} - |u|^{-\alpha}q^{\alpha-1} \quad .$$

If $\alpha \neq 0$ a little calculation shows that it is equal to $\pi(v)\Gamma(\pi^{-1}) + \pi^{-1}(v)\Gamma(\pi)$. If $\alpha = 0$ we get

$$(1-q^{-1})(-m+1) - q^{-1} - q^{-1} = (1-q^{-1})(\log_q(\frac{1}{|uv|})+1) - 2/q \ .$$

Now suppose that $|uv| = q^m$, $m > 1$. Since $|uv| > 1$,

$$J_\pi(u,v) = \pi^{-1}(u)P\int_{|x|\leq 1}\bar{\chi}(uv/x)\pi(x)|x|^{-1}dx$$

$$+ \pi^{-1}(u)\int_{1<|x|<|uv|} \bar{\chi}(uv/x)\bar{\chi}(x)\pi(x)|x|^{-1}dx$$

$$+ \pi^{-1}(u)P\int_{|x|\geq|uv|} \bar{\chi}(x)\pi(x)|x|^{-1}dx$$

$$= \pi(v) P \int_{|x| \geq |uv|} \bar{\chi}(x) \pi^{-1}(x) |x|^{-1} dx + \pi^{-1}(u) P \int_{|x| \geq |uv|} \bar{\chi}(x) \pi(x) |x|^{-1} dx$$

$$+ \pi^{-1}(u) \sum_{1 \leq k < m} F_{\pi}(k, uv) \quad .$$

If π is unramified we see that the first two terms are zero (by (5.2)) and the third (by (5.23)(i)) is $F_{\pi}(m/2, uv)$ if m is even and is zero otherwise.

If π is ramified of degree h, $h \geq 1$, consider first the case $1 < m \leq h$. From (5.23) (ii) the third term is zero and from the definition of the gamma function the first two terms are $\pi(v)\Gamma(\pi^{-1})$ and $\pi^{-1}(u)\Gamma(\pi)$ respectively. If $m > h$ then by (5.2) the first two terms are zero and the evaluation of the third term is established by reference to (5.23)(ii).

Remark. $J_{\pi}(u,v)$ can be extended in obvious ways to characters π that are not unitary to obtain Bessel functions of more general order.

Remark. The proof of (5.25) shows that for π umramified, $J_{\pi}(u,v) \rightarrow J_1(u,v)$ as $\pi \rightarrow 1$ and that

(5.26) $$|J_{\pi}(u,v)| \leq \begin{cases} (1-q^{-1})(\log_q \dfrac{1}{|uv|} + 1) + 2/q, & |uv| \leq q \\ (1-q^{-1}) & , |uv| > q \end{cases} .$$

<u>Corollary (5.27)</u>. (i) <u>For</u> $\pi \in \hat{K}^*$, $\pi \neq 1$, $J_\pi(u,v)$ <u>is bounded as a function of</u> $u,v \in K^*$. $J_1(u,v)$ <u>is bounded for</u> (uv) <u>bounded away from zero</u>, $J_1(u,v) \simeq \log_q(\frac{1}{|uv|})(1-q^{-1})$ <u>as</u> $|uv| \to 0$.

(ii) <u>For fixed</u> $u,v \in K^*$, $J_\pi(u,v)$ <u>is bounded as a function of</u> $\pi \in \hat{K}^*$

<u>Proof</u>. (i) If $\pi \neq 1$ the result follows from (5.24) and the fact that $|\pi(u)| = |\pi(v)| = 1$. For $\pi = 1$ the result is contained in (5.26). The asymptotic formula for $J_1(u,v)$ is immediate from (5.25)(i)(b).

(ii) Fix (u,v). If π is umramified then

$$|J_\pi(u,v)| \leq \begin{cases} (1-q^{-1})[\log_q \dfrac{1}{|uv|} + 1] + 2/q, & |uv| \leq q \\ (1-q^{-1}) & , |uv| > q \end{cases} .$$

If π is ramified then,

$$\begin{aligned} |J_\pi(u,v)| &\leq \max[|\Gamma(\pi)| + |\Gamma(\pi^{-1})|, 3(1-q^{-1})] \\ &\leq \max[2\, q^{-\frac{1}{2}}, 3(1-q^{-1})]. \end{aligned}$$

The next three theorems assert that $J_\pi(u,v)$ can be regarded as the Mellin transform of $\bar{\chi}(ux + v/x)$, the Fourier transform of $P\,\bar{\chi}(v/x)\pi(x)|x|^{-1}$ and as the inverse Mellin transform of $(\pi\pi')^{-1}\Gamma^{\#}(\pi')\Gamma^{\#}(\pi\pi')$.

Theorem (5.28). Let $f(x) = \bar{\chi}(ux + v/x)$, $u,v \in K^*$. Then $f \in \mathscr{S}^{*\prime}$, $\mathfrak{M} f(\pi) = J_\pi(u,v) \in \hat{\mathscr{S}}^{*\prime}$.

Proof. $\bar{\chi}(ux + v/x)$ is bounded as a function of x. From (5.27) we see that $J_\pi(u,v)$ is bounded as a function of π. Thus they are distributions in $\mathscr{S}^{*\prime}$ and $\hat{\mathscr{S}}^{*\prime}$ respectively. For $\varphi \in \hat{\mathscr{S}}^*$ we wish to show that $\int_{K^*} \bar{\chi}(ux + v/x)(\mathfrak{M}^{-1}\varphi)^\sim(x)|x|^{-1}dx = \int_{\hat{K}^*} J_\pi(u,v)\varphi(\pi)d\pi$.

We fix $u,v \in K^*$, $\varphi \in \hat{\mathscr{S}}^*$. Since φ is compactly supported we may fix h_0 such that $\varphi(\pi) = 0$ if $\deg(\pi) > h_0$.

Choose a compact set $C = C(u,v,h_0,\varphi) \subset K^*$ so that

$$J_\pi(u,v) = \int_C \bar{\chi}(ux + v/x)\pi(x)|x|^{-1}dx ,$$

and C contains the support of $(\mathfrak{M}^{-1}\varphi)^\sim$. Then

$$\int_{\hat{K}^*} J_\pi(u,v)\varphi(\pi)d\pi = \int_{\pi \in A'_{h_0}} \varphi(\pi)\int_C \bar{\chi}(ux + v/x)\pi(x)|x|^{-1}dx\, d\pi$$

$$= \int_C \bar{\chi}(ux + v/x) \int_{\pi \in A'_{h_0}} \varphi(\pi)\pi(x)d\pi|x|^{-1}dx$$

$$= \int_C \bar{\chi}(ux + v/x)(\mathfrak{M}^{-1}\varphi)^\sim(x)|x|^{-1}dx$$

$$= \int_{K^*} \bar{\chi}(ux + v/x)(\mathfrak{M}^{-1}\varphi)^\sim(x)|x|^{-1}dx .$$

<u>Lemma (5.29)</u>. $P\ \bar{\chi}(v/x)\pi(x)|x|^{-1} = f \in \mathscr{S}'$, <u>as a function of</u> $x \in K^+$. $\hat{f}(u) = J_\pi(u,v) \in \mathscr{S}'$.

<u>Proof</u>. Fix $\pi \in \hat{K}^*$, $v \in K^*$. $J_\pi(u,v)$ is locally integrable on K^+ as a function of u, and so it is in $\mathscr{S}'$. We want to show next that $f = P\ \bar{\chi}(v/x)\pi(x)|x|^{-1} \in \mathscr{S}'$. Let $h = \max[1,\deg(\pi)]$. Fix $\varphi \in \mathscr{S}$, and choose a positive integer k such that φ is constant on cosets of $\mathfrak{P}^k$ and is supported on $\mathfrak{P}^{-k}$. Now choose ℓ so $q^\ell = \max[q^h|v|^{-1}, q^k]$. If $n > \ell$ then

$$(5.30) \qquad \int_K [\bar{\chi}(v/x)\pi(x)|x|^{-1}]_n \varphi(x)\,dx$$

$$= \varphi(0) \int_{q^{-n} \le |x| < q^{-\ell}} \bar{\chi}(v/x)\pi(x)|x|^{-1}dx + \int_{q^{-\ell} \le |x| \le q^k} \varphi(x)\bar{\chi}(v/x)\pi(x)|x|^{-1}dx$$

$$= \int_{q^{-\ell} \le |x| \le q^k} \varphi(x)\bar{\chi}(v/x)\pi(x)|x|^{-1}dx\ .$$

Thus, $P \int \bar{\chi}(v/x)\pi(x)|x|^{-1}\varphi(x)\,dx$

$$= \int_{q^{-\ell} \le |x| \le q^k} \bar{\chi}(v/x)\pi(x)|x|^{-1}\varphi(x)\,dx\ ,$$

where k and ℓ depend on π, v and φ. For $\{\varphi_s\}$ a null sequence in $\mathscr{S}$ we may choose a fixed pair k and ℓ and it is obvious that $f \in \mathscr{S}'$, and that $[\bar{\chi}(v/x)\pi(x)|x|^{-1}]_n \to f$.

For all n large enough we have

$$(\hat{f},\varphi) = (f,\hat{\varphi}) = \int_K [\bar{\chi}(v/x)\pi(x)|x|^{-1}]_n \hat{\varphi}(x)dx$$

$$= \int_K \{[\bar{\chi}(v/\cdot)\pi(\cdot)|\cdot|^{-1}]_n\}^{\wedge}(u)\varphi(u)du$$

$$= \int_K \varphi(u) \int_{q^{-n}\leq|x|\leq q^n} \bar{\chi}(ux+v/x)\pi(x)|x|^{-1}dx\ du \quad ,$$

If $\pi \not\equiv 1$ the inner integral converges boundedly to $J_\pi(u,v)$ as $n \to \infty$. By the Legesgue dominated convergence theorem we obtain $(\hat{f},\varphi) = \int_K \varphi(u)J_\pi(u,v)du$, so $\hat{f}(u) = J_\pi(u,v)$ if $\pi \not\equiv 1$.

The rest of the proof is concerned with the single case $\pi(x) \equiv 1$. Without loss we may assume that $|v| \leq q^{k+1}$

$$(\hat{f},\varphi) = \varphi(0) \int_{|u|\leq q^{-k}} \int_{q^{-n}\leq|x|\leq q^n} \bar{\chi}(ux + v/x)|x|^{-1}dx\ du$$

$$+ \int_{q^{-k}\leq|u|\leq q^k} \varphi(u) \int_{q^{-n}<|x|\leq q^n} \bar{\chi}(ux + vx)|x|^{-1}dx\ du \quad .$$

In the second integral $|u|$ is bounded away from zero, v is fixed, so the inner integral converges boundedly to $J_1(u,v)$.

To complete the proof we need to show that if $|v| \leq q^{k+1}$

$$\lim_{n\to\infty} \int_{|u|\leq q^{-k}} \int_{q^{-n}\leq|x|\leq q^n} \bar{\chi}(ux + v/x)|x|^{-1}dx\ du = \int_{|u|\leq q^{-k}} J_1(u,v)du \quad .$$

Observe that $|uv| \leq q$. If $n > k$ we may write

$$\int_{|u|\leq q^{-k}} \int_{q^{-n}\leq |x|\leq q^{n}} \bar{\chi}(ux + v/x) \frac{dx}{|x|} du$$

$$= \int_{|u|\leq q^{-k}} \left[\int_{q^{-n}\leq |x|\leq q^{k}} \bar{\chi}(v/x) \frac{dx}{|x|} + \int_{q^{k}<|x|\leq q^{n}} \bar{\chi}(ux) \frac{dx}{|x|} \right] du$$

$$= \int_{|u|\leq q^{-k}} \int_{q^{-k}|v|\leq |x|\leq q} \bar{\chi}(x) \frac{dx}{|x|} du + \int_{|u|\leq q^{-n}} \int_{q^{k}<|x|\leq q^{n}} \frac{dx}{|x|} du$$

$$+ \int_{q^{-n}<|u|<q^{-k}} \int_{q^{k}<|x|\leq q|u|^{-1}} \bar{\chi}(ux) \frac{dx}{|x|} du$$

$$= I_1 + I_2 + I_3 .$$

$$I_2 = (n-k)q^{-n}(1-q^{-1}) \to 0 \quad \text{as} \quad n \to \infty .$$

Similarly $I_3 \to \int_{|u|\leq q^{-k}} \int_{q^{k}<|x|\leq q|u|^{-1}} \bar{\chi}(ux) \frac{dx}{|x|} du$, and so

$$I_1 + I_2 + I_3 \to \int_{|u|\leq q^{-k}} \left\{ \int_{q^{-k}|v|\leq|x|\leq q} + \int_{q^{k}|u|<|x|\leq q} \bar{\chi}(x) \frac{dx}{|x|} \right\} du .$$

The inner integrals may be evaluated directly. If $|v| = q^{a}$, $|u| = q^{-b}$ so $|uv| = q^{a-b}$ and $\log_q 1/|uv| = b-a$ we obtain that the inner integral is equal to $[(k-a+1)(1-q^{-1})-q^{-1}] - [(b-k)(1-q^{-1})-q^{-1}$ $[(b-a) + 1] - 2/q = [\log_q \frac{1}{|uv|} + 1](1/q^{-1}) - 2/q = J_1(u,v)$.

Theorem (5.31). Let $f(v) = J_\pi(u,v)$. Then $f \in \mathscr{S}^{*\prime}$ and

$$\mathfrak{M}f(\pi') = (\pi\pi')^{-1}(u)\Gamma^{\#}(\pi')\Gamma^{\#}(\pi\pi') .$$

We omit the proof. It is harder than the proof of (5.19), but similar in manner.

Definition. For $\pi \in \hat{K}^*$, $\varphi \in \mathscr{S}$ we define the Hankel transform (of order π) of φ by

(5.32) $$H_\pi\varphi(v) = \int_K \varphi(u)J_\pi(u,v)du \quad , \; v \in K^* .$$

Remarks. H_π is well defined on $\mathscr{S}$ since $J_\pi(u,v)$ is locally integrable. If $\pi \not\equiv 1$ then $J_\pi(u,v)$ is bounded as a function of v so H_π has a well defined extension to $f \in L^1(K^+)$.

(5.33) $$\left.\begin{array}{l} H_\pi f(v) = \int f(u)J_\pi(u,v)du \in L^\infty(K) \\ \|H_\pi f\|_\infty \le A_\pi\|f\|_1 \end{array}\right\} f \in L^1 ,$$

where $A_\pi = \|J_\pi(u,v)\|_\infty$, $\pi \not\equiv 1$.

Lemma (5.34). If $\varphi \in \mathscr{S}$ then $H_\pi\varphi(v) = (\pi^{-1}(\cdot)|\cdot|^{-1}\hat{\varphi}(1/\cdot))^\wedge(v)$, $H_\pi\varphi \in L^2(K^+)$ and $\|H_\pi\varphi\|_2 = \|\varphi\|_2$.

Proof. By (5.29) if $f = P(\overline{\chi}(v/x)\pi^{-1}(x)|x|^{-1})$ then $\hat{f} \in \hat{\mathscr{S}}'$ and $\hat{f} = J_\pi(\cdot,v)$. Thus,

$$H_\pi\varphi(v) = (J_\pi(\cdot,v),\varphi) = (f,\hat{\varphi}) = P\int_K \bar{\chi}(v/x)\pi(x)|x|^{-1}\hat{\varphi}(x)dx$$

$$= P\int_K \bar{\chi}(vx)\pi^{-1}(x)|x|^{-1}\hat{\varphi}(1/x)dx.$$

$$\|\pi^{-1}(\cdot)|\cdot|^{-1}\hat{\varphi}(1/\cdot)\|_2^2 = \int|x|^{-2}|\hat{\varphi}(1/x)|^2dx = \int |\hat{\varphi}(x)|^2dx$$

$$= \|\varphi\|_2^2 .$$

Thus, $\pi^{-1}(x)|x|^{-1}\hat{\varphi}(1/x) \in L^2$, and by Plancherel ((2.6) and (2.3))

$$H_\pi\varphi(v) = P\int_K \bar{\chi}(vx)\pi^{-1}|x|^{-1}\hat{\varphi}(1/x)dx = (\pi^{-1}(\cdot)|\cdot|^{-1}\hat{\varphi}(1/\cdot))\hat{}\,(v),$$

$$H_\pi\varphi \in L^2(K),\ \|H_\pi\varphi\|_2 = \|(\pi^{-1}(\cdot)|\cdot|^{-1}\hat{\varphi}(1/\cdot))\hat{}\,\|_2 = \|\varphi\|_2 .$$

__Corollary (5.35)__. __If__ $\hat{\varphi} \in \mathscr{S}^*$ (__equivalently__ $\varphi \in \mathscr{S}, \int \varphi = 0$) __then__ $H_\pi\varphi \in \mathscr{S}$.

__Proof__. $\hat{\varphi}(x) \in \mathscr{S}^*$ iff $\hat{\varphi}(1/x) \in \mathscr{S}^*$. Thus $\hat{\varphi} \in \mathscr{S}^* \Longrightarrow$ $\pi^{-1}(x)|x|^{-1}\hat{\varphi}(1/x) \in \mathscr{S}^* \subset \mathscr{S} \Longrightarrow H_\pi\varphi = (\pi^{-1}(\cdot)|\cdot|^{-1}\hat{\varphi}(1/\cdot))\hat{} \in \mathscr{S}.$

__Theorem (5.36)__. __If__ $\pi \in \hat{K}^*$, $\pi \not\equiv 1$, __then__ H_π __is a bounded map from__ $L^1(K^+)$ __into__ $L^\infty(K^+)$. __If__ $f \in L^1$, $H_\pi f(v)$ __is continuous for__ $v \neq 0$. $H_\pi f(v) \to 0$ __as__ $|v| \to \infty$.

$$H_\pi f(v) = \Gamma(\pi)\int \pi^{-1}(u)f(u)du + \Gamma(\pi^{-1})\pi(v)\int f(u)du + o(1) \text{ as } v \to 0 .$$

__Proof__. The first part of the theorem follows from the remarks above (5.32).

Since $J_\pi(u,v)$ is continuous in $v \in K^*$ for each $u \in K^*$ and $|J_\pi(u,v)f(u)| \leq A_\pi|f(u)| \in L^1$, the Lebesgue dominated convergence theorem shows that $H_\pi f(v)$ is continuous for $v \neq 0$.

If $k = \max[1,\deg(\pi)]$ and $|u| \leq q^k|v|^{-1}$ we have that $J_\pi(u,v) = \Gamma(\pi)\pi^{-1}(u) + \Gamma(\pi^{-1})\pi(v)$. Thus,

$$H_\pi f(v) = \Gamma(\pi) \int_{|u|\leq q^k|v|^{-1}} \pi^{-1}(u)f(u)\,du + \Gamma(\pi^{-1})\pi(v) \int_{|u|\leq q^k|v|^{-1}} f(u)\,du$$

$$+ \int_{|u|> q^k|v|^{-1}} O(1)f(u)\,du$$

$$= \Gamma(\pi)\int \pi^{-1}(u)f(u)\,du + \Gamma(\pi^{-1})\pi(v)\int f(u)\,du + o(1), \ |v| \to 0 .$$

From the fact that $\mathscr{S}$ is dense in L^1 and $f \to H_\pi f$ is bounded from $L^1 \to L^\infty$ it will suffice to show that $H_\pi\varphi(v) \to 0$ as $|v| \to \infty$ for $\varphi \in \mathscr{S}$. We assume that φ is constant on cosets of $\mathfrak{P}^n$ and is supported on $\mathfrak{P}^{-n}$ for some $n \geq 1$. Then $\hat{\varphi}$ has the same property. Hence $\hat{\varphi}(1/x)$ is zero if $|x| < q^{-n}$ and $\hat{\varphi}(1/x) = \varphi(0)$ if $|x| \geq q^n$. Thus,

$$H_\pi\varphi(v) = \int_{q^{-n}\leq |x|< q^n} \pi^{-1}(x)|x|^{-1}\hat{\varphi}(1/x)\overline{\chi}(vx)\,dx + \varphi(0)\int_{|x|\geq q^{-n}} \pi^{-1}(x)|x|^{-1}\overline{\chi}(vx)\,dx$$

$$= \int_{q^{-n}\leq |x|< q^n} \pi^{-1}(x)|x|^{-1}\hat{\varphi}(1/x)\overline{\chi}(vx)\,dx,$$

for $|v|$ large enough ($|v| \geq q^{h-n+1}$, $h = \deg(\pi)$).

On the set $\{q^{-n} \leq |x| < q^n\}$, $\hat{\varphi}(1/x)$ is constant on cosets of $\mathfrak{P}^{3n}$ and π^{-1} is constant on cosets of $\mathfrak{P}^{n+h+1}$ where $h = \deg(\pi)$. Set $k = \max[3n, n+h+1]$ and write $\{q^{-n} \leq |x| < q^n\}$ as a union of cosets of $\mathfrak{P}^k$. If $|v| > q^k$ it now follows that $H_\pi\varphi(v) = 0$. That is, if $\varphi \in \mathcal{S}$, $H_\pi\varphi$ has compact support in K^+, and a fortiori, $H_\pi\varphi(v) \to 0$ as $|v| \to \infty$.

Theorem (5.37). H_π *has a unique extension to* $L^2(K^+)$ *as a unitary operator.* *If* $f,g \in L^2(K^+)$ *then*

(a) $\int H_\pi f(x)g(x)\,dx = \int f(x)H_{\pi^{-1}}g(x)\,dx$

(b) $\int H_\pi f(x)\overline{g(x)}\,dx = \int f(x)\pi(-1)\overline{H_\pi g(x)}\,dx$.

Thus, $\pi(-1)H_\pi$ *is the adjoint of* H_π *and so is its inverse.*

Proof. Clearly H_π has a unique extension to L^2 as a linear isometry, as follows from (5.34). If (a) is established then (b) follows since by a trivial change of variables argument $H_{\pi^{-1}}\bar{g}$ $\pi(-1)\overline{H_\pi g}$. (One changes variables in (5.32) for $g \in \mathcal{S}$ and then takes limits in L^2.) Furthermore, if (a) holds we can show that H_π is onto and hence is unitary. For if H_π is not onto there is a $g \in L^2$, $g \not\equiv 0$ such that for all $f \in L^2$, $0 = \int H_\pi f(x)g(x)\,dx = \int f(x)H_{\pi^{-1}}g(x)\,dx$. Thus, $H_{\pi^{-1}}g \equiv 0$. But this implies that $g \equiv 0$, since $H_{\pi^{-1}}$ is an isometry. A contradiction.

To prove (a) we compute formally:

$$\int H_\pi f(v)g(v)dv = \int (\pi^{-1}(\cdot)|\cdot|^{-1}\hat{f}(1/\cdot))^\wedge(v)g(v)dv$$

$$= \int \pi^{-1}(x)|x|^{-1}\hat{f}(1/x)\hat{g}(x)dx$$

$$= \int \hat{f}(x)\pi(x)|x|^{-1}\hat{g}(1/x)dx$$

$$= \int f(v)(\pi(\cdot)|\cdot|^{-1}\hat{g}(1/\cdot))^\wedge(v)dv = \int f(v)H_{\pi^{-1}}g(v)dv.$$

All these relations hold if f, g, $H_\pi f$ and $H_{\pi^{-1}}g \in \mathscr{S}$. This follows from (5.35) if $\hat{f},\hat{g} \in \mathscr{S}^*$. Since the set of all such f,g are dense in L^2 the result follows by taking limits.

6. Some elementary aspects of Fourier series on the ring of integers

Throughout this section, K is a fixed local field, $\mathfrak{O}$ is the ring of integers in K, $\mathfrak{P}$ is the prime ideal in $\mathfrak{O}$, $\mathfrak{O}/\mathfrak{P} \cong GF(q) = \mathfrak{k}$, $q = p^c$, p a prime, c a positive integer, and $\rho : \mathfrak{O} \to \mathfrak{k}$ the canonical homonorphism of $\mathfrak{O}$ onto $\mathfrak{k}$.

Our object of study is harmonic analysis on the group (ring) $\mathfrak{O}$. If f is a measurable function on $\mathfrak{O}$ we might simply consider f as defined on $\mathfrak{O}$ (and nowhere else) or we might identify f with its extension $\mathcal{E}f(x) = f(x)$ if $x \in \mathfrak{O}$, $\mathcal{E}f(x) = 0$ if $x \notin \mathfrak{O}$. Similarly we may take any function g that is measurable on K and consider its restriction to $\mathfrak{O}$ or perhaps the function defined on K by $\mathfrak{R}g = g\Phi_0$.

Any of these views may be taken as is convenient. An example of the possible confusion that may arise is given by the following: Let χ be a character on K^+. Since $\mathfrak{D}$ is a subgroup of K^+, the restriction of χ to $\mathfrak{D}$ is a character on $\mathfrak{D}$ and we call it χ. At the same time we need the function $\chi_{p^{-k}}$ whose values are

$\chi_{p^{-k}}(x) = \chi(p^{-k}x)$ for $x \in \mathfrak{D}$ where χ is not the "restriction" but the "original" character on $\mathfrak{D}$. We shall unashamedly use "χ" for both.

Our first job is to identify the characters on $\mathfrak{D}$ (which are countable in number) and find a "natural" order for them.

We observe, as above, that if χ_u is any character on K^+ then $\chi_u|_{\mathfrak{D}} \equiv \chi_u$ is a character on $\mathfrak{D}$, since $\mathfrak{D}$ is a subgroup of K^+. Now we note that, as characters on $\mathfrak{D}$, $\chi_u = \chi_v$ iff $u-v \in \mathfrak{D}$, which is to say that $\chi_u = \chi_v$ if $u + \mathfrak{D} = v + \mathfrak{D}$ and $\chi_u \neq \chi_v$ if $u+\mathfrak{D} \cap v+\mathfrak{D} = \emptyset$. To see this we note that $\chi_u(x) = \chi_v(x)$ for all $x \in \mathfrak{D}$ iff $\chi((u-v)x) \equiv 1$ for all $x \in \mathfrak{D}$ which is true iff that χ is trivial on the sphere of radius $|u-v|$, which is true iff $u - v \in \mathfrak{D}$.

Consequently, if $\{u(n)\}_{n=0}^{\infty}$ is a complete list of (distinct) coset representatives of $\mathfrak{D}$ in K^+ then $\{\chi_{u(n)}\}_{n=0}^{\infty}$ is a list of distinct characters on $\mathfrak{D}$. There are several ways of seeing that this list of characters is complete. For those with an abstract orientatio notice that $\hat{K}^+ = K^+$, and that the annihalator of $\mathfrak{D}$ in $\hat{K}^+$ is $\mathfrak{D}$ an hence $\hat{\mathfrak{D}} \cong \hat{K}^{+}/\mathfrak{D} = K/\mathfrak{D}$. Operating more constructively: Let X be a character on $\mathfrak{D}$ that is not on the list. Then identify X with its extension to K (setting it equal to zero outside of $\mathfrak{D}$). Then $c_n = \int_K X(x)\overline{x}_{u(n)}(x)dx = 0$ for all n. Observe that $X \in L^1(K)$ and $\hat{X}(\xi) = c_n$ if $\xi \in u(n) + \mathfrak{D}$. Thus, $\hat{X} \equiv 0$ and so $X \equiv 0$, which is impossible since X is a character on $\mathfrak{D}$.

We obtain:

<u>Proposition (6.1).</u> <u>Let</u> $\{u(n)\}_{n=0}^{\infty}$ <u>be a complete list of (distinct) coset representatives of</u> $\mathfrak{D}$ <u>in</u> K^{+}. <u>Then</u> $\{\chi_{u(n)}|_{\mathfrak{D}}\}_{n=0}^{\infty}$ <u>is a complete list of (distinct) characters on</u> $\mathfrak{D}$. <u>Furthermore it is a complete orthonormal system on</u> $\mathfrak{D}$.

When such a list is obtained we define <u>Fourier coefficents</u> $\{c_n\}_{n=0}^{\infty}$, $c_n = \int_{x \in \mathfrak{D}} f(x)\overline{\chi}_{u(n)}(x)dx$, $f \in L^1(\mathfrak{D})$, and write $f(x) \sim \sum_{n=0}^{\infty} c_n \chi_{u(n)}(x)$, in the usual manner. The series is called the Fourier series of f. This is formally equivalent to considering f as a function on $L^1(K)$ (set equal to zero outside of $\mathfrak{D}$ supported on $\mathfrak{D}$ and hence with a Fourier transform, $\hat{f}$, that is constant on cosets of $\mathfrak{D}$, $\hat{f}(u(n)) = c_n$. The collections $\{c_n\} = \{\hat{f}(u(n))\}$ will be identified and both are referred to as the Fourier coefficients of f.

The standard L^2 theory for compact abelian groups shows that the Fourier series of f converges to f in $L^2(\mathfrak{D})$ and that $\int_{\mathfrak{D}}|f|^2 = \sum_{n=0}^{\infty}|c_n|^2$. These results hold without regard to the ordering of the characters. We now proceed to impose a "natural" order on the sequence $\{u(n)\}_{n=0}^{\infty}$.

Note that $\mathfrak{k} = GF(q)$ is a c-dimensional vector space over $GF(p) \subset \mathfrak{k}$. We choose a set $\{1 = \epsilon_0, \epsilon_1, \ldots, \epsilon_{c-1}\} \subset \mathfrak{D}^*$ such that $\{\rho(\epsilon_k)\}_{k=0}^{c-1}$ is a basis of $GF(q)$ over $GF(p)$.

For n, $0 \leq n < q$, $n = a_0 + a_1 p + \cdots + a_{c-1}p^{c-1}$, $0 \leq a_k < p$, $k = 0,1,\ldots,c-1$ we define

$$(6.2) \qquad u(n) = (a_0 + a_1\epsilon_1 + \cdots + a_{c-1}\epsilon_{c-1})\mathfrak{p}^{-1}, \quad 0 \leq n < q .$$

Note that $u(0) = 0$, $\{u(n)\}_{n=0}^{q-1}$ is a complete set of coset representatives of $\mathfrak{D}$ in $\mathfrak{P}^{-1}$ and $|u(n)| = q$ if $1 \leq n < q$. Now write $n = b_0 + b_1 q + \cdots + b_s q^s$, $0 \leq b_k < q$, $n \geq 0$, and set

$$(6.3) \qquad u(n) = u(b_0) + \mathfrak{p}^{-1}u(b_1) + \mathfrak{p}^{-2}u(b_2) + \cdots + \mathfrak{p}^{-s}u(b_s).$$

Note that for $n, m \geq 0$ it is not true that $u(n+m) = u(n) + u(m)$. However, it is true that for all $r \geq 0$, $k \geq 0$ $u(rq^k) = \mathfrak{p}^{-k}u(r)$, and for $r \geq 0$, $k \geq 0$, $0 \leq t < q^k$, $u(rq^k + t) = u(rq^k) + u(t) = \mathfrak{p}^{-k}u(r) + u(t)$.

Let us now write $\chi_n = \chi_{u(n)}|_{\mathfrak{D}}$, $n \geq 0$. We have,

<u>Proposition (6.4).</u> $\{\chi_n\}_{n=0}^{\infty}$ <u>is a complete set of characters on</u> $\mathfrak{D}$. <u>For</u> $r \geq 0$, $k \geq 0$, $0 \leq t < q^k$, $\chi_{r \cdot q^k + t} = \chi_{r \cdot q^k} \chi_t$ <u>and</u> $\chi_{rq^k}(x) = \chi_r(\mathfrak{p}^{-k}x)$.

<u>Definition.</u> (a) The Dirichlet kernels are the functions

$$D_n(x) = \sum_{k=0}^{n-1} \chi_k(x),\ n \geq 1. \quad D_0(x) \equiv 0 \text{ by convention.}$$

(b) If $f \in L^1(\mathfrak{D})$ the Fourier coefficients $\{c_n\}_{n=0}^{\infty} = \{\hat{f}(u(n)\}_{n=0}^{\infty}$ are given by $c_n = \int_{\mathfrak{D}} f(x)\bar{\chi}_n(x)dx$. The Fourier series is given by

$$f(x) \approx \sum_{n=0}^{\infty} c_n \chi_n(x).$$

(c) The $n^{\underline{th}}$ partial sum of the Fourier series of f is denoted $S_n f(x)$ and is defined as, $S_n f(x) = \sum_{\nu=0}^{n-1} c_\nu \chi_\nu(x)$.

<u>Remark.</u> Various results that obtain immediately from viewing f as a function defined on K, but supported on $\mathfrak{D}$ will be taken for granted. Thus: If $f \in L^1(\mathfrak{D})$, $f \approx \sum c_n \chi_n$ then $c_n = o(1)$ as $n \longrightarrow \infty$; If $f \in L^1(\mathfrak{D})$ and $c_n \equiv 0$ then $f \equiv 0$; If $f \in L^2(\mathfrak{D})$ then $\int |f|^2 = \sum |c_n|^2$.

<u>Proposition (6.5)</u>. <u>If</u> $f \in L^1(\mathfrak{O})$ <u>then</u>

$$S_n f(x) = \int f(\xi) D_n(x-\xi)\, d\xi = \int f(x-\xi) D_n(\xi)\, d\xi \quad .$$

<u>Proof</u>. Left as an exercise. The argument is standard.

<u>Remark</u>. $S_n f = f * D_n$ where convolution can be defined in either of two ways. We can restrict the functions to $\mathfrak{O}$, or extend them to K and take the convolution on K, and some trivial arithmetic shows that value of such a convolution is zero off of $\mathfrak{O}$. In either case if $f,g \in L^1$ we see that $(f * g)(x) \sim \sum \hat{f}(u(n))\hat{g}(u(n))\chi_n(x)$.

<u>Proposition (6.6)</u>. $u(n) = 0$ <u>iff</u> $n = 0$. If $k \geq 1$ <u>then</u> $|u(n)| = q^k$ <u>iff</u> $q^{k-1} \leq n < q^k$. <u>Consequently</u> $D_n(x)$ <u>is constant on cosets of</u> $\mathfrak{P}^k$ <u>iff</u> $n \leq q^k$.

<u>Lemma (6.7)</u>. <u>If</u> $f \in L^1(\mathfrak{O})$, $x \in \mathfrak{O}$, $k \geq 0$ <u>then</u>

$$S_{q^k} f(x) = q^k \int_{|x-\xi| \leq q^{-k}} f(\xi)\, d\xi = \frac{1}{|x+\mathfrak{P}^k|} \int_{x+\mathfrak{P}^k} f(\xi)\, d\xi$$

<u>as follows from the fact that</u> $D_{q^k} = q^k \Phi_k$.

<u>Proof.</u> We need to show that $D_{q^k}(x) = q^k \Phi_k(x), x \in \mathfrak{O}$. The Fourier coefficients of D_{q^k} are $\{c_\nu\}$ where $c_\nu = 1$, $0 \leq \nu < q^k$, and 0 if $\nu \geq q^k$. From (6.6) it follows that if we view D_{q^k} as a function on $\mathfrak{O}$, then $\hat{D}_{q^k} = \Phi_{-k}$. Thus $D_{q^k} = q^k \Phi_k$.

__Corollary (6.8)__. (a) __If__ $f \in L^1(\mathfrak{D})$, $S_{q^k}f(x) \to f(x)$ a.e.. __In particular__, $S_{q^k}f(x) \to f(x)$ __at each point where__ f __is continuous__.

(b) __If__ $f \in L^p(\mathfrak{D})$, $1 \le p < \infty$, $S_{q^k}f \to f$ __in__ L^p.

(c) __If__ f __is continuous on__ $\mathfrak{D}$, $S_{q^k}f \to f$ __uniformly__.

__Proof__. (a) is a reformulation of (1.15). To prove (b) and (c) note that $S_{q^k}f(x) - f(x) = q^k \int_{|x-z| \le q^{-k}} (f(x-z) - f(x))dz.$

From Minkowski's integral inequality we have

$$\|S_{q^k}f(\cdot) - f(\cdot)\|_p \le q^k \int_{|z| \le q^{-k}} \|f(\cdot - z) - f(\cdot)\|_p \, dz .$$

For $f \in L^p$, $1 \le p < \infty$; f continuous, $p = \infty$ respectively, we have $\|f(\cdot - z) - f(\cdot)\|_p = o(1)$ as $|z| \to 0$. Thus, $\|S_{q^k}f(\cdot) - f(\cdot)\|_p \to 0$ as $k \to \infty$.

We list some properties of $D_n(x)$:

(6.9) $\int D_n = 1$ __for all__ $n \ge 1$. Obvious.

(6.10) $|D_n(x)| \le n$ __for all__ $n \ge 0$. Obvious.

(6.11) $D_{r \cdot q^k + t}(x) = D_{q^k}(x) D_r(\mathfrak{p}^{-k}x) + \chi_r(\mathfrak{p}^{-k}x) D_t(x)$,

$r \ge 0$, $k \ge 0$, $0 \le t < q^k$.

<u>Proof</u>. $$D_{r\cdot q^k+t}(x) = \sum_{\rho=0}^{r-1}\sum_{s=0}^{q^k-1} \chi_{\rho q^k+s}(x) + \sum_{s=0}^{t-1} \chi_{rq^k+s}(x)$$

$$= \sum_{\rho=0}^{r-1} \chi_{\rho q^k}(x) \sum_{s=0}^{q^k-1} \chi_s(x) + \chi_{rq^k}(x) \sum_{s=0}^{t-1} \chi_s(x)$$

$$= D_r(\mathfrak{p}^{-k}x)D_{q^k}(x) + \chi_r(\mathfrak{p}^{-k}x)D_t(x) .$$

(6.12) $|D_n(x)| \leq q|x|^{-1}, \ x \neq 0, \ x \in \mathfrak{O}.$

<u>Proof</u>. Let $|x| = q^{-k+1}$. Write $n = r\cdot q^k + t, \ 0 \leq t < q^k$ as in (6.11

Since $|x| = q^{-k+1}$, $D_{q^k}(x) = 0$ so $D_n(x) = \chi_r(\mathfrak{p}^{-k}x)D_t(x)$. By (6.10)

$|D_t(x)| \leq t$, so $|D_n(x)| \leq t < q^k = q|x|^{-1}$.

(6.13) $D_{q^k}(x) = \prod_{s=0}^{k-1} D_q(\mathfrak{p}^{-s}x)$, $k \geq 1$.

<u>Proof</u>. For $k = 1$ this is an identity. To complete the proof we need to show that $D_{q^{k+1}}(x) = D_q(\mathfrak{p}^{-k}x)D_{q^k}(x), \ k \geq 0$. But this follows from (6.11) with $r = q$ and $t = 0$.

<u>Theorem (6.14)</u>. (A "Dini" convergence test) <u>If</u> $f \in L^1(\mathfrak{O})$ <u>and</u> $\int |f(x_0-z) - f(x_0)| \, |z|^{-1}dz < \infty$ <u>then</u> $S_nf(x_0) \rightarrow f(x_0)$ <u>as</u> $n \rightarrow \infty$.

<u>Proof</u>. Fix $\epsilon > 0$. Then choose k so that

$\int_{\mathfrak{p}^k} |f(x_0-z) - f(x_0)| \, |z|^{-1}dz < \epsilon/2q$. Then let $g_\epsilon(z) = (f(x_0-z)-f(x_0))(1-\Phi_k$

Then

$$|S_n f(x_0) - f(x_0)| = \left|\int (f(x_0-z) - f(x_0))D_n(z)dz\right|$$

$$\leq q \int_{\mathfrak{P}^k} |f(x_0-z) - f(x_0)|\ |z|^{-1} dz + \left|\int g_\epsilon(z)\, D_n(z)dz\right| .$$

By our choice of k, the first term is bounded by $\epsilon/2$. Since the support of g_ϵ is on the complement of $\mathfrak{P}^k$ it follows from (6.11) (writing $n = r \cdot q^k + t$) that $D_n(z) = \sum_{s=0}^{t-1} \chi_{r\cdot q^k+s}(z)$ for $|z| > q^{-k}$, $0 \leq t < q^k$. Hence the second term is bounded by the sum of at most q^k Fourier coefficients all of whose orders are greater than or equal to $n\,q^{-1}$. As $n \to \infty$ this sum tends to zero. Thus, for n large enough $|S_n f(x_0) - f(x_0)| < \epsilon$.

<u>Corollary (6.15)</u>. <u>If</u> $f \in L^1(\mathfrak{O})$ <u>and</u> f <u>is constant on</u> $x_0 + \mathfrak{P}^k$ <u>for some</u> k <u>then</u> $S_n f(x_0) \to f(x_0)$. <u>In particular, if f is radial</u> (i.e., $f(x) = a_k$, $|x| = q^{-k}$, $k \geq 0$) <u>then</u> $S_n f(x) \to f(x)$ <u>for all</u> $x \neq 0$.

<u>Remark</u>. From our construction of the sequence of characters we see that if $n = b_0 + b_1 q + \cdots + b_s q^s$ then $\chi_n = \chi_{\mathfrak{p}^{-1}b_0} \chi_{\mathfrak{p}^{-2}b_1} \chi_{\mathfrak{p}^{-3}b_2} \cdots \chi_{\mathfrak{p}^{-s+1}b_s}$. Suppose $q = p$ is a prime. Then $0 \leq b_k \leq p-1$ and $\chi_{\mathfrak{p}^{-s-1}b_s} = \chi_{b_s p^s} = (\chi_{p^s})^{b_s}$ and $\chi_{p^s}(x) = \chi_1(\mathfrak{p}^{-s}x)$. Thus, in the case that $q = p$ is a prime, $\{\chi_{p^s}\}_{s=0}^{\infty}$ is the precise analogue of the Rademacher

functions. All other characters χ_n are obtained by the rule:
$\chi_n = \prod_{k=0}^{s} (\chi_{p^k})^{b_k}$, $0 \leq b_k < p$, $n = b_0 + b_1 p + \cdots + b_s p^s$.

<u>Definition</u>. $\mathcal{S} = \mathcal{S}(\mathfrak{O})$ is the collection of <u>test functions on $\mathfrak{O}$</u>. $\varphi \in \mathcal{S}$ iff φ is constant on cosets of $\mathfrak{P}^k$ in $\mathfrak{O}$ for some $k \geq 0$.

Viewed as a function with support $\mathfrak{O} = \mathfrak{P}^0$, $\varphi \in \mathcal{S}(K)$, φ is constant on cosets of $\mathfrak{P}^k$ so $\hat{\varphi}$ is constant on cosets of $\mathfrak{O}$ and is supported on $\mathfrak{P}^{-k}$. That is, $\hat{\varphi}(u(n)) = 0$ if $|u(n)| > q^k$ and so φ is a "polynomial". That is,

$$\varphi(x) = \sum_{|u(n)| \leq q^k} \hat{\varphi}(u(n))\chi_n(x) = \sum_{n=0}^{q^k - 1} \hat{\varphi}(u(n))\chi_n(x) = S_{q^k}\varphi(x) = S_m \varphi(x), \ m \geq$$

A null sequence in $\mathcal{S}(\mathfrak{O})$ is a sequence $\{\varphi_s\}$ of functions in $\mathcal{S}$ that are all constant on the same $\mathfrak{P}^k$ and tend uniformly to zero. This is easily seen to be equivalent to the requirement that $\hat{\varphi}_s(u(n))$ if $|u(n)| > q^k$ for some fixed k and $\hat{\varphi}_s(u(n)) \rightarrow 0$ as $s \rightarrow \infty$ f each n. $\mathcal{S}(\mathfrak{O})$ takes the topology, as a topological vector space, induced by this definition of a null sequence.

<u>Definition</u>. $\mathcal{S}' = \mathcal{S}'(\mathfrak{O})$ is the <u>space of distribution on $\mathfrak{O}$.</u> It is the topological dual of $\mathcal{S}$ and is given the Weak* topology.

We denote the action of $f \in \mathcal{S}'(\mathfrak{O})$ on $\varphi \in \mathcal{S}(\mathfrak{O})$ by $(f,\varphi)_{\mathfrak{O}}$ or by $(f,\varphi$ if there is little chance of confusion. We will occasionally use $(f,$ when

$$f \in \mathcal{S}'(K), \ \varphi \in \mathcal{S}(K).$$

If $f \in \mathscr{S}'$ then the set of values $\{(f,\chi_n)\}_{n=0}^{\infty}$ are known since $\chi_n \in \mathscr{S}$. But then $(f,\varphi) = \sum (f,\chi_n)\hat{\varphi}(u(n))$, and in this sense $f = \sum (f,\bar{\chi}_n)\chi_n = \sum (f,\chi_n)\bar{\chi}_n$. (Note that $\{\bar{\chi}_n\} = \{\chi_{-u(n)}|_{\mathfrak{D}}\}$ is also a complete set of characters. That is, the series $\sum (f,\bar{\chi}_n)\chi_n$ converges in $\mathscr{S}'$ to $f \in \mathscr{S}'$. Furthermore, any complex sequence $\{d_n\}$ can occur as the values of $(f,\chi_n) = d_n$. Thus, $\mathscr{S}'$ is in 1 : 1 correspondence with the collection of complex sequences $\{d_n\}_{n=0}^{\infty}$.

If $f \in \mathscr{S}'(K)$ and $\psi \in \mathscr{S}$ we can define $f\psi = \psi f \in \mathscr{S}'$ by the rule $(f\psi,\varphi) = (f,\psi\varphi)$, for all $\varphi \in \mathscr{S}$. Thus, if $f \in \mathscr{S}'$, $f\Phi_0 \in \mathscr{S}'(K)$ is well defined and may also be interpreted as an element of $\mathscr{S}'(\mathfrak{D})$ by viewing $\varphi \in \mathscr{S}(\mathfrak{D})$ as if it was defined on K but supported on $\mathfrak{D}$.

With the convention as stated we want to find the Fourier series of $f\Phi_0$ for $f \in \mathscr{S}'(\mathfrak{D})$ in terms of the Fourier transform of f, $\hat{f}$.

$$(f\Phi_0,\bar{\chi}_n)_{\mathfrak{D}} = (f,\Phi_0\,\chi_{-u(n)})_K\,.$$

From (1.2) we get $(\tau_{u(n)}\Phi_0)^{\wedge} = \chi_{-u(n)}\hat{\Phi}_0 = \chi_{-u(n)}\Phi_0$. Thus,

$$(f\Phi_0,\bar{\chi}_n)_{\mathfrak{D}} = (f,(\tau_{u(n)}\Phi_0)^{\wedge})_K = (\hat{f},\tau_{u(n)}\Phi_0)_K\,.$$

That is, $(f\Phi_0,\bar{\chi}_n)$, the Fourier coefficient of $f\Phi_0$ is the Fourier transform of f applied to the characteristic function of $u(n)+\mathfrak{D}$, which, if $\hat{f}$ were a function, would be

$$(f\Phi_0,\bar{\chi}_n)_{\mathfrak{D}} = \int_{u(n)+\mathfrak{D}} \hat{f}(z)\,dz = \hat{f}\star\Phi_0(u(n)).$$

In the next chapter we will justify interpreting $(f, \tau_x \varphi) = f * \varphi$ for $f \in \mathcal{S}'$, $\varphi \in \mathcal{S}$, but for now we accept it as a notational convenie
We conclude:

Theorem (6.16). *If* $f \in \mathcal{S}'(K)$ *then* $f|_{\mathfrak{D}} \in \mathcal{S}'(\mathfrak{D})$ *and* $f|_{\mathfrak{D}} = \sum (\hat{f} * \Phi_0)(u(n)$

There are particular instances of special interest.

Corollary (6.17). *Suppose* $f \in \mathcal{S}'$, $f\Phi_0 \in L^1$ *and* $\hat{f}$ *is a function,* $u \in K$. *Then*

(a) $$\int_{|x|\le 1} f(x)\overline{\chi}_u(x)\,dx = \int_{|x|\le 1} \hat{f}(x+u)\,dx$$

(b) $$\int_{|x|\le 1} f(x)\,dx = \int_{|x|\le 1} \hat{f}(x)\,dx \quad .$$

Notice that the right hand side of (b) is the average of $\hat{f}$ over the integers, so that what one has here is a local field version of t Poisson summation formula.

Some examples. We have shown that $(|x|^{\alpha-1})^\wedge = \Gamma_1(\alpha)|x|^{-\alpha}$. Let us suppose that $0 < \mathrm{Re}(\alpha) < 1$. Note that (6.17)(b) gives us another evaluation of $\Gamma_1(\alpha)$; namely $\Gamma_1(\alpha) = \int_{|x|\le 1} |x|^{\alpha-1}dx / \int_{|x|\le 1} |x|^{-\alpha}dx$.

If we let $e_\alpha = \int_{|x|\le 1} |x|^{\alpha-1}dx = (1-q^{-1})/(1-q^{-\alpha})$

$$|x|^{\alpha-1}|_{\mathfrak{D}} = e_\alpha + \Gamma_1(\alpha)\sum_{\nu\neq 0} |u(\nu)|^{-\alpha}\chi_\nu(x) \quad (\mathcal{S}') \ .$$

Since $|x|^{\alpha-1}$ is radial it follows from (6.15) that equality holds pointwise for $x \neq 0$.

As another example, suppose π^* is a character on $\mathfrak{O}^*$ that is ramified of degree 1. Then $\widehat{\pi^*}(x) = \Gamma_{\pi^*}(1)\overline{\pi^*}(x)|x|^{-1}$. Thus, as distributions, and as functions (for $x \neq 0$) we have,

$$\pi^*(x)|_{\mathfrak{O}} = c_{\pi^*}q^{\frac{1}{2}}\sum_{\nu=1}^{\infty}\overline{\pi^*}(u(\nu))|u(\nu)|^{-1}\chi_{\nu}(x).$$

<u>Definition.</u> The <u>arithmetic means, Ceasaro sums,</u> or <u>(C,1) means</u> of a series $\sum_{\nu=0}^{\infty} a_\nu$, are defined as follows: We let $s_n = \sum_{\nu=0}^{n-1} a_\nu$ be the ordinary partial sums. Let $\sigma_n = (s_1 + s_2 \cdots + s_n)/n$. We call $\{\sigma_n\}_{n=1}^{\infty}$ the (C,1) means of the series and say that the series is (C,1) summable (to s) ($\sum a_\nu = s$ (C,1)) if $\sigma_n \to s$ as $n \to \infty$.

It is relatively easy to see that if $\sum a_\nu = s$ (i.e., $s_n \to s$) then the series is (C,1) summable to the same sum. If the sequence $\{a_\nu\}_{\nu=0}^{\infty}$ is sufficiently regular we get a converse.

<u>Lemma (6.18)</u>. (a) <u>If</u> $s_n \to s$ <u>then</u> $\sigma_n \to s$.

(b) <u>If</u> $a_\nu = O(1/\nu)$ <u>and</u> $\sigma_n \to s$ <u>then</u> $s_n \to s$.

These are basic results in summability theory.

If we look back to (6.8) an alternative to (C,1) summability is suggested. Consider, instead of $\{\sigma_n\}$ the series of q^kth means $\{s_{q^k}\}$. For if $s_n \to s$ then so does $s_{q^k} \to s$, since it is a subsequence. Moreover we can show that if $\{a_\nu\} = o(1/\nu)$ we get a converse. Thus, write $n = q^k + r$ where $0 \le r < (q-1)q^k$. Then

$$s_n = s_{q^k} + \sum_{\nu=q^k}^{q^k+r-1} a_\nu \, , \; |s_n - s_{q^k}| \le \sum_{\nu=q^k}^{q^{k+r-1}} |a_\nu|$$

$$\le q^{k+1}\Big(\sup_{|\nu| \ge q^k} |a_\nu|\Big) = o(1) \quad \text{as} \quad n \to \infty \, ,$$

since $k \to \infty$ as $n \to \infty$. If we try to strengthen this by considerin a condition $a_\nu = O(1/\nu)$ we do not get a converse. Consider the sequence $\{1, 2^{-1}, 2^{-2}, -2^{-2}, 2^{-3}, 2^{-3}, -2^{-3}, -2^{-3}, \cdots\}$, and $q = 2$. $s_{2^k} = 3$ for $k \ge 1$. But $s_{2^k + 2^{k-1}} = 7/4$ for $k \ge 1$ so $\lim s_n$ does not exi but $|a_\nu| \le 1/(1+\nu)$ for all ν.

It is not possible to avoid the issue and we will consider the properties of (C,1) sums of Fourier series. (6.20) below will illustrate importance of the $O(1/\nu)$ condition.

Definition. Let f be a bounded function on $\mathfrak{D}$. For $k \ge 0$ let $\{S_\nu^k\}_{\nu=1}^{q^k}$ be a listing of the cosets of $\mathfrak{P}^k$ in $\mathfrak{D}$. We say that f i of <u>bounded fluctuation</u> in $\mathfrak{D}$ if

$$V^*f = \sup_{k \geq 0} \sum_{\nu=1}^{q^k} \sup_{x,y \in S_\nu^k} |f(x) - f(y)| < \infty .$$

The <u>total variation of f</u>, Vf is defined by

$$V f = \lim \sup_k \Big(\sum_{\nu=1}^{q^k} \sup_{x,y \in S_\nu^k} |f(x) - f(y)| \Big) .$$

<u>Theorem (6.19)</u>. <u>If</u> f <u>is of bounded fluctuation then</u> f <u>is continuous for</u> $x \notin E$, <u>where</u> E <u>is countable</u>.

<u>Proof</u>. Let $\operatorname{osc} f(x) = \lim_{k \geq 0} \sup \sup_{y \in \mathfrak{P}^k} |f(x+y) - f(x)|$. f is continuous at x iff $\operatorname{osc} f(x) = 0$. Let $E_0 = \{x \in \mathfrak{O}: \operatorname{osc} f(x) > 1\}$. Let $F \subset E_0$ be finite. Clearly $\sum_{x \in F} \operatorname{osc} f(x) \leq V^*f$. Thus $\operatorname{card}(F) < V^*f$ and so E_0 is a finite set. Similarly, $E_k = \{x \in \mathfrak{O} : 2^{-k} < \operatorname{osc} f(x) \leq 2^{-k+1}\}$, $k = 1,2,\cdots$ are also finite sets, so $E = \cup_k E_k$ is countable.

<u>Theorem (6.20)</u>. <u>Suppose</u> f <u>is of bounded fluctuation and</u> $f(x) \sim \sum c_n \chi_n(x)$. <u>Then</u> $|n\, c_n| \leq q\, V^*f$, $n \geq 1$ <u>and</u> $\lim \sup_n |n\, c_n| \leq q\, V f$. <u>In particular</u>, $c_n = O(1/n)$.

<u>Proof</u>. Suppose $n > 0$, $|u(n)| = q^k$, $k \geq 1$. Then $q^{k-1} \leq n < q^k$. Note that $\int_{S_\nu^{k-1}} \overline{\chi}_n(x)dx = 0$ for all ν. Fix a set of elements $\{a_\nu^k\}$ such that $S_\nu^k = a_\nu^k + \mathfrak{P}^k$. Then

$$c_n = \int f(x)\,\bar{\chi}_n(x)\,dx = \sum_{\nu=1}^{q^{k-1}} \int_{S_\nu^{k-1}} f(x)\,\bar{\chi}_n(x)$$

$$= \sum_{\nu=1}^{q^{k-1}} \int_{S_\nu^{k-1}} (f(x) - f(a_\nu^{k-1}))\bar{\chi}_n(x)\,dx \quad .$$

Thus, $|c_n| \leq \left[\sum_{\nu=1}^{q^{k-1}} \sup_{x,y \in S_\nu^{k-1}} |f(x) - f(y)| \right] \cdot q^{-k+1}$

$$\leq q\,|n|\,V^* f \; .$$

Similarly, $\limsup |n\,c_n| \leq q\; Vf$.

Example. Recall the example discussed following (6.17). $\pi^*|_{\mathfrak{D}}$, wher[e] $\deg(\pi^*) = 1$. $\pi^*|_{\mathfrak{D}}$ is of bounded fluctuation. It is easy to check th[at] $V(\pi^*|_{\mathfrak{D}}) = 2$. But $\pi^*|_{\mathfrak{D}}(x) \sim \sum c_n \chi_n(x)$ with $|c_n| = q^{\frac{1}{2}}|u(n)|^{-1}$.

Thus, $|n\,c_n| \leq q^{\frac{1}{2}} \leq 2 \cdot q = q\;Vf$, but $|n\,c_n| \geq q^{-\frac{1}{2}}$.

Definition. The Fejer kernels, $K_n(x)$, $n \geq 0$ are the functions $K_n(x) = (1/n)\sum_{\nu=1}^{n} D_\nu(x) = \sum_{\nu=0}^{n-1}(1-\nu/n)\chi_\nu(x)$, $n \geq 1$, $K_0(x) \equiv 0$.

Let $\sigma_n f(x) = (1/n)\sum_{\nu=1}^{n} S_\nu f(x) = \sum_{\nu=0}^{n-1}(1-\nu/n)c_\nu\chi_\nu(x)$. Then

$\sigma_n f(x) = \int f(x-\xi)K_n(\xi)d\xi = \int f(\xi)K_n(x-\xi)d\xi$, are the (C,1) means of th[e] partial sums of the Fourier series of f, whenever $f \in \mathcal{S}'(\mathfrak{D})$.

(6.21) $\quad |K_n(x)| \le (n+1)/2, \quad n \ge 1$

(6.22) $\quad \int K_n(x)\,dx = 1, \quad n \ge 1$

(6.23) $\quad |K_n(x)| \le q|x|^{-1}, \quad n \ge 1$

(6.24) $\quad \int |K_n(x)|\,dx \le A$, $\quad A > 0$ independent of n.

(6.25) $\quad$ For each $k \ge 0$ $\quad \int_{q^{-k} \le |x| \le 1} |K_n(x)|\,dx = o(1)$ as $n \to \infty$.

(6.21) is obvious since $|D_n(x)| \le n$ and so

$$|K_n(x)| \le (1/n)\sum_{\nu=1}^{n} |D_\nu(x)| \le (1/n)\sum_{\nu=1}^{n} \nu = n(n+1)/2n = (n+1)/2 .$$

For (6.22), $\int K_n = (1/n)\sum_{\nu=1}^{n} \int D_\nu = (1/n)(\sum_{\nu=1}^{n} 1) = 1$. Similarly, each $|D_\nu(x)| \le q|x|^{-1}$, so $|K_n(x)| \le q|x|^{-1}$. (6.24) and (6.25) are considerably harder and we postpone the proofs. We assume them for the purpose of establishing the next three results.

<u>Theorem (6.26)</u>. (a) <u>If</u> $f \in L^p(\mathfrak{D})$, $1 \le p < \infty$, $\sigma_n f \to f$ <u>in</u> L^p .

(b) <u>If</u> f <u>is continuous</u>, $\sigma_n f \to f$ <u>uniformly</u>.

(c) <u>If</u> f <u>is a bounded function and</u> f <u>is continuous at</u> x_0 <u>then</u> $\sigma_n f(x_0) \to f(x_0)$.

Proof. $\sigma_n f(x) - f(x) = \int (f(x-\xi) - f(x))K_n(\xi)d\xi$. Thus,

$\|\sigma_n f(\cdot) - f(\cdot)\|_p \leq \int \|f(\cdot - \xi) - f(\cdot)\|_p K_n(\xi)d\xi$, where for

(a) $f \in L^p$, $1 \leq p < \infty$ and for (b) f is continuous and in either case $\|f(\cdot - \xi) - f(\cdot)\|_p = o(1)$ as $|\xi| \to 0$ and $\|f(\cdot - \xi) - f(\cdot)\|_p \leq 2 \|f\|_p$.

Fix $\epsilon > 0$ and choose $k(\epsilon) = k$ so $\|f(\cdot - \xi) - f(\cdot)\|_p < \epsilon/2A$ for $|\xi| \leq q^{-k}$ where $A > 0$ is the constant of (6.24). Then,

$$\|\sigma_n f - f\|_p \leq (\epsilon/2A) \int_{|\xi|\leq q^{-k}} |K_n(\xi)|d\xi + 2\|f\|_p \int_{q^{-k}<|\xi|\leq 1} |K_n(\xi)|d\xi$$

$$\leq (\epsilon/2A)\int |K_n(\xi)|d\xi + 2\|f\|_p\ o(1) \qquad \text{(by (6.25))}$$

$$\leq \epsilon/2 + o(1) \quad \text{as} \quad k \to \infty . \qquad \text{(by (6.24))}$$

Thus, $\|\sigma_n f - f\|_p = o(1)$ as $n \to \infty$ and the proof of (a) and (b) is complete.

For (c) note that f is continuous at x_0 so we may choose $k = k(\epsilon)$ so $|f(x_0 - \xi) - f(x_0)| < \epsilon/2A$ if $|\xi| \leq q^{-k}$ and proceed as above.

Corollary (6.27). *If f is of bounded fluctuation then $\sigma_n f(x) \to f(x)$ for all x where f is continuous, and hence for all but a countable set.*

Proof. f is of bounded fluctuation implies that f is bounded and is continuous except for $x \in D$, E a countable set (6.19). By (6.20)

$f(x) \approx \sum c_n \chi_n(x)$ where $c_n = O(1/n)$. By (6.26) $\sigma_n f(x) \to f(x)$ if $x \notin E$. By (6.18) $S_n f(x) \to f(x)$ if $x \notin E$.

Corollary (6.28). If f is a continuous function of bounded fluctuation then $S_n f \to f$ uniformly.

Proof. The proof depends upon certain details in the proof of (6.18). Namely, the rate of convergence of σ_n to s when $s_n \to s$ and $a_\nu = O(1/\nu)$ depends only on the rate of convergence of s_n to s and the constant B that occurs in the relation $|a_\nu| \leq B|\nu|^{-1}$. But from (6.26)(b) we see that $\sigma_n f(x) \to f(x)$ uniformly. From (6.20) we see that $|c_\nu \chi_\nu(x)| \leq |c_\nu| \leq q\, V^* f|\nu|^{-1}$ uniformly in x, so $S_n f(x) \to f(x)$ uniformly.

We now proceed to establish (6.24) and (6.25).

(6.29) Let $n = r q^k + t$, $r \geq 0$, $k \geq 0$, $0 \leq t < q^k$, then

$$n K_n(x) = q^k D_{q^k}(x)\, r\, K_r(\mathfrak{p}^{-k}x) + t\, D_{q^k}(x) D_r(\mathfrak{p}^{-k}x)$$

$$+ D_r(\mathfrak{p}^{-k}x) q^k K_{q^k}(x) + \chi_r(\mathfrak{p}^{-k}x)\, t\, K_t(x).$$

Proof. Recall that $D_{\rho q^k + \tau}(x) = D_{q^k}(x) D_\rho(\mathfrak{p}^{-k}x) + \chi_\rho(\mathfrak{p}^{-k}x) D_\tau(x)$. (See (6.11).) Then write,

$$n\,K_n(x) = \sum_{\rho=0}^{r-1} \sum_{\tau=1}^{q^k} D_{\rho q^k+\tau}(x) + \sum_{\tau=1}^{t} D_{rq^k+\tau}(x) ,$$

expand and add.

Let us now assume that $q^{\ell} \le n < q^{\ell+1}$, $\ell > k$, $|x| = q^{-k+1}$. (Note that $q^{\ell-k} \le n\,q^{-k} < q^{\ell-k+1}$; $n\,q^{-k} = r + t\,q^{-k}$, so $q^{\ell-k-1} \le r \le q^{\ell-k+1}$ as $n \to \infty$).

Since $|x| > q^{-1}$, $D_{q^k}(x) = 0$ so

$$\begin{aligned} n\,K_n(x) &= D_r(\mathfrak{p}^{-k}x)\,q^k K_{q^k}(x) + \chi_r(\mathfrak{p}^{-k}x)\,t\,K_t(x) \\ &= n\,I_1^n(x) + n\,I_2^n(x) . \end{aligned}$$

$|I_2^n(x)| \le (t/n)\,|K_t(x)| \le t^2/n < q^2/n|x|^2$, since $t^2 \le (q^k)^2 = q^2|x|^{-2}$. Thus

$$\begin{aligned} \int_{|x|=q^{-k+1}} |I_2^n(x)|\,dx &\le (A/n) \int_{|x|=q^{-k+1}} |x|^{-2}dx = a\,q^k/n \\ &\le A\,q^{k-\ell} . \end{aligned}$$

<u>Note</u>. In these arguments A is a positive constant that is independent of n and k, but can change its value, even in the same set of inequalities.

Similarly $|I_1^n(x)| \le (q^{2k}/n)|D_r(\mathfrak{p}^{-k}x)|$, $|x| = q^{-k+1}$.

Thus, $$\int_{|x|=q^{-k+1}} |I_1^n(x)|\,dx \leq (q^{2k}/n) \int_{|x|=q^{-k+1}} |D_r(\mathfrak{p}^{-k}x)|\,dx$$

$$= ((q^{2k}q^{-k})/n) \int_{|x|=1} |D_r(\mathfrak{p}^{-1}x)|\,dx$$

$$\leq (q^k/n) \quad_{|x|\leq 1} |D_r(\mathfrak{p}^{-1}x)|\,dx$$

$$\leq (q^k/n)\left[\int_{|x|\leq 1} |D_r(\mathfrak{p}^{-1}x)|^2 dx\right]^{\frac{1}{2}}$$

$$\leq (q^k r^{\frac{1}{2}})/n = (r\cdot q^k/n)r^{-\frac{1}{2}} \leq A\, q^{(k-\ell)/2} \quad .$$

If we note that $\ell > k$, we see that the two estimates for $I_1^n(x)$ and $I_2^n(x)$ may be assembled to obtain:

<u>Lemma (6.30)</u>. <u>Let</u> $q^\ell \leq n < q^{\ell+1}$, $|x| = q^{-k+1}$, $1 \leq k \leq \ell-1$. <u>Then</u> <u>there is a constant</u> $A > 0$ <u>independent of</u> n <u>and</u> k <u>such that</u>

$$\int_{|x|=q^{-k+1}} |K_n(x)|\,dx \leq A\, q^{(k-\ell)/2} \quad .$$

We again set n, $q^\ell \leq n < q^{\ell+1}$. For all x, $|K_n(x)| \leq n < q^{\ell+1}$.

Hence $$\int_{|x|< q^{-\ell+1}} |K_n(x)|\,dx \leq q^{\ell+1}\cdot q^{-\ell+1} = q^2 = A.$$

We now use (6.30).

$$\int_{|x|\geq q^{-\ell+2}} |K_n(x)|\,dx = \sum_{k=1}^{\ell-1} \int_{|x|=q^{-k+1}} |K_n(x)|\,dx$$

$$\leq A \sum_{k=1}^{\ell-1} q^{(k-\ell)/2} = A \sum_{k=1}^{\ell-1} q^{-k/2} \leq A(1-q^{-\frac{1}{2}})^{-1} = A.$$

This establishes (6.24).

Note that $\ell = \ell(n) \to \infty$ as $n \to \infty$.

$$\int_{q^{-k}\leq |x|\leq 1} |K_n(x)|\,dx = \sum_{\nu=1}^{k+1} \int_{|x|= q^{-\nu+1}} |K_n(x)|\,dx$$

$$\leq A \sum_{\nu=1}^{k+1} q^{(\nu-\ell)/2} \leq A\, q^{(k-\ell)/2} = o(1) \quad \text{as } n\to\infty$$

This establishes (6.25).

A problem of general interest in Fourier series is the relation between the "smoothness" of a function and the "size" of its Fourier coefficients. For example, one such result is (f of bounded fluctuati $\Longrightarrow$ ($c_n = O(1/n)$). Another example is the Riemann-Lebesgue lemma; namely that $f \in L^1 \Longrightarrow c_n = o(1/n)$. That this last result is of the particular type is shown by the following consideration:

Let $\omega_p(q^{-k};f) = \omega_p(q^{-k}) = \sup_{0\leq |h|\leq q^{-k}} \|f(\cdot+h) - f(\cdot)\|_p$.

Then if $f \in L^p(\mathfrak{D})$, $1 \le p < \infty$ then $\omega_p(q^{-k};f) = o(1)$, $k \to \infty$.

If f is continuous then $\omega_\infty(q^{-k};f) = o(1)$, $k \to \infty$.

Proposition (6.31). *If* $f \in L^1(\mathfrak{D})$, $f(x) \sim \sum c_n \chi_n(x)$, *then* $|c_n| \le \omega_1(q|u(n)|^{-1};f)$ *as* $n \to \infty$.

Proof. Take $n > 0$. Then there is an h, $|hu(n)| = q$ such that $|\chi_n(h)-1| \ge 1$. Furthermore $\chi_n(h)c_n = \int f(x+h)\bar{\chi}_n(x)dx$. Thus,

$(\chi_n(h)-1)c_n = \int (f(x+h) - f(x))\bar{\chi}_n(x)dx$, or

$$|c_n| \le (1/|\chi_n(h)-1|) \int |f(x+h) - f(x)|dx$$

$$\le \omega_1(|h|;f) = \omega_1(q|u(n)|^{-1};f).$$

Another example are the so-called Lipschitz or Hölder conditions.

Definition. Let $\alpha > 0$. We say that $f \in \Lambda_\alpha(\mathfrak{D}) = \Lambda_\alpha = \text{Lip}\,\alpha$, if there is a constant $A > 0$ such that $|f(x)-f(y)| \le A|x-y|^\alpha$ for all $x,y \in \mathfrak{D}$.

If $f \in \Lambda_\alpha$ it is easy to see that $\omega_1(q^{-k};f) \le \omega_\infty(q^{-k};f) = O(q^{-k\alpha})$ as $k \to \infty$.

Corollary (6.32). *If* $f \in \Lambda_\alpha(\mathfrak{D}), f(x) \sim \sum c_n \chi_n(x)$ *then* $c_n = O(n^{-\alpha})$ *as* $n \to \infty$.

In point of fact, the Plancherel equality is another example, since it says that if the function is in L^2 its coefficients decrease fast enough to be in ℓ^2.

Proposition (6.33) (Plancherel). If $f \in L^2(\mathfrak{D})$, $f(x) \approx \sum c_n \chi_n(x)$ then $\sum |c_n|^2 = \int |f|^2$.

The next results lie somewhat deeper.

Theorem (6.34) (Bernstein's Theorem). If $f \in \Lambda_\alpha(\mathfrak{D})$, $\alpha > 1/2$, $f(x) \approx \sum c_n \chi_n(x)$ then $\sum |c_n| < \infty$.

Proof. $\sum |c_n| = |c_0| + \sum_{k=1}^{\infty} \sum_{|u(n)|=q^k} |c_n| \leq |c_0| + A \sum_{k=1}^{\infty} q^{k/2} [\sum_{|u(n)|=q^k} |c_n|$

Fix k, $k \geq 1$. Take $h = 0$. Then

$$\omega_2^2(|h|;f) \geq \int |f(x+h)-f(x)|^2 dx = \sum_n |\chi_n(h)-1|^2 |c_n|^2 \geq \sum_{|u(n)|=q^k} |\chi_n(h)-1|^2 |c$$

Now suppose $|u(n)| = q^k$. Then there is an h, $|h| = q^{-k+1}$ such that $|\chi_n(h)-1| > 1$, since on the $(q-1)$ cosets of $\mathfrak{P}^k$ in $\mathfrak{P}^{k-1} \sim \mathfrak{P}^k$, χ_n takes on at least one value with a negative real part. Summing over these $(q-1)$ cosets we see that $(q-1)\omega_2^2(q^{-k+1};f) \geq \sum_{|u(n)|=q^k} |c_n|^2$.

Thus, $\sum |c_n| \leq c_0 + A \sum_{k=1}^{\infty} q^{k/2} \omega_2(q^{-k+1};f)$.

If $f \in \Lambda_\alpha$, then $\omega_2(q^{-k+1};f) \le \omega_\infty(q^{-k+1};f) = O(q^{-k\alpha})$. Thus, $\sum |c_n| \le c_0 + A\sum_{k=1}^{\infty} q^{k(\frac{1}{2}-\alpha)}$. If $\alpha > 1/2$ the series converges.

Theorem (6.35). If f is of bounded fluctuation and $f \in \Lambda_\alpha$, $\alpha > 0$ then $\sum |c_n| < \infty$, for $f(x) \approx \sum c_n \chi_n(x)$.

Proof. Let V^*f be as in the definition of bounded fluctuation. Let $\{a_\nu^k\}$ be the q^k coset representatives of $\mathfrak{B}^k$ in $\mathfrak{D}$, as in the proof of (6.20). Then for each $x \in \mathfrak{D}$,

$$\sum_{\nu=1}^{q^{k-1}} |f(x+a_\nu^{k-1}+h) - f(x+a_\nu^{k-1})|^2 \le \omega_\infty(|h|;f)V^*f .$$

If we integrate the left hand side with respect to x,

$$q^{k-1}\int |f(x+h) - f(x)|^2dx \le \omega_\infty(|h|;f)V^*f .$$

Arguing as in (6.34) we have

$$\sum_{|u(n)|=q^k} |\chi_n(h)-1|^2|c_n|^2 \le \int |f(x+h) - f(x)|^2dx \le q^{-k+1}\omega_\infty(q^{-k+1};f)V^*f .$$

Thus, $\left[\sum_{|u(n)|=q^k}|c_n|^2\right]^{\frac{1}{2}} \le A\, q^{-k/2}\omega_\infty^{\frac{1}{2}}(q^{-k+1};f)(V^*f)^{\frac{1}{2}}$.

If $f \in \Lambda_\alpha$, $\omega_\infty(q^{-k+1};f) = O(q^{-k\alpha})$, and so $\left[\sum_{|u(n)|=q^k}|c_n|^2\right]^{\frac{1}{2}} = O(q^{-k(\alpha+\frac{1}{2})})$. As before we see that $\sum |c_n| < \infty$.

Example. The following example is instructive. It is an example of a continuous function of bounded fluctuation that does not have an absolutely convergent Fourier series. Let $f(x) = \pi(x)/\log_q(1/|x|)$. A straight forward calculation shows that $f(x) \approx \Gamma(\pi) \sum_{n>0} c_n \chi_n(x)$,

with $c_n = \dfrac{\pi^{-1}(u(n))}{(\frac{\lfloor u(n) \rfloor}{q})\log_q(\frac{\lfloor u(n) \rfloor}{q})}$. We have assumed that $\deg(\pi) = 1$.

If $\deg(\pi) = h > 1$ then the series starts with $n = q^h$.

7. Notes for Chapter II

General references for §1 - §5 are Gelfand-Graev [1] and Sally and Taibleson [1]; for §6 the only reference available is an announcement of results for the p-series case, Taibleson [2].

The approach we take throughout these lecture notes is that all functions are complex-valued. For an idea of the scope of the problem if one dealt with local field-valued functions see the survey article by Monna [1] or the results of Schikhof [1] on non-archimedean harmonic analysis.

§1. In this section we develop many results explicitly that we could have obtained directly from the theory of locally compact abelian groups. Examples are (1.1), (1.2), (1.6), (1.7) and (1.8). A good general reference is Hewitt and Ross [1].

The linear space $\mathscr{S}$, is discussed in more detail in §3, where it is introduced as the space of test functions for the space of distributions on K.

The dilations introduced in (1.5) are dilations by the elements $\mathfrak{p}^k$. Later we look at delations by any non-zero t in K. (See §5, the definition preceeding (5.8) and the subsequent discussion of homogeneity through (5.10).)

(1.12) is the local field version of Norbert Wiener's covering lemma. See Stein and Weiss [1, p.54, Lemma (3.3)] for a modern restatement of Wiener's covering lemma.

Once (1.14) is established the powerful results on inversion of the Fourier transform, (1.15) - (1.18) follow as in the euclidean case.

(1.19) and (1.20) should be compared to the corresponding results for Gauss-Weierstrass kernels and Poisson kernels. See, for example, Stein and Weiss [1,p.6:(1.13) and (1.14), p.16:(1.27) and (1.28)]. This explains the use of the symbol "A_k" in (1.15) which is our substitute for Abel summability.

§2. We develop here the usual extensions of results for the Fourier transform on L^1 to the transform on L^2 by use of the fact that $L^1 \cap L^2$ is dense in L^2.

In the proof of (2.1) we use the fact that if $f \in L^2(K)$ then $\|f(\cdot + y) - f(\cdot)\|_2 = o(1)$ as $y \to 0$ in K. The most general fact of this nature we will need is the following:

Lemma. Let G be a locally compact abelian group and let $f \in L^p(G)$, $1 \leq p < \infty$. Then $\|f(\cdot + y) - f(\cdot)\|_p = o(1)$ as $y \to 0$ in G.

The proof is easy, following from the density of the class of compactly supported continuous functions in $L^p(G)$.

§3. This is the usual distribution theory. Its principal virtue is that it allows us to extend the Fourier transform to all the L^p spaces.

A useful heuristic device: If $\varphi \in \mathcal{S}$, then the larger the cosets on which φ is constant, the "smoother" it is; the smaller its support the "faster it decreases at infinity". Viewed in this way, (3.2) describes the classical situation in which the local smoothness of a function determines the rate of decrease at infinity of its Fourier transform and conversely.

The crucial results in this section are (3.9), (3.10) and the definitions which lead to them.

Distributions will be considered further in III §3, where we study harmonic analysis on K^n. We note that every statement or theorem in §1 - §3 of this chapter could also be stated for K^n (with only slight notational changes). We could have started directly on K^n, but for the usual pedagogical reasons we started with the one-dimensional case. Furthermore, we now have all the tools we need to work out the details of several important applications (in §5, §6) as soon as we take a brief look at analysis on the multiplicative structure, K^*, in §4.

§4. We are concerned here with a few elementary aspects of harmonic analysis on K^*. Our approach will be to take over, wholesale, the elementary theory of locally compact abelian groups, mostly without explicit reference.

In (4.1) we use the notation dx^*. dx^* is a Haar measure on $K^* = K \sim \{0\}$ as a multiplicative group. We choose $dx/|x| = dx^*$. For a Haar measure on $(K^*)^\wedge$ we use the notation $d\pi$. We determine $d\pi$ up to a constant, a, which we determine (by means of Plancherel's theorem) later in the section.

It will be instructive for the reader to compare (4.5) and (3.2).

§5. The results of this section are, essentially, the results in Sally and Taibleson [1]; namely, the explicit calculation of the gamma, beta and Bessel functions and the introduction of the Hankel transform as a unitary operator on L^2. The gamma and beta functions were introduced in Gelfand and Graev [1].

Following the proof of (5.5) we commented: "We now show that the gamma function arises naturally as the normalizing factor of a certain distribution." The discussion which follows leads to Theorems (5.6) an (5.7), the statement of (5.7) being taken by Gelfand and Graev [1] as their definition of the gamma function.

(5.7) also shows that our gamma function is essentially the invari factor $\rho(\pi)$ defined by Tate (see Lang [1,VII §3]) in his study of the local zeta function. Tate considered the class of test functions $\{f: |x|^{\sigma} f \in L^1, |x|^{\sigma} \hat{f} \in L^1 \text{ for all } \sigma \geq 0\}$. Clearly, any function in is in this class, but it is strictly larger than $\mathcal{S}$ as we see by considering the function $\varphi(x) = \begin{cases} 0, & |x| \leq 1 \\ e^{-|x|}, & |x| > 1 \end{cases}$.

For $0 < \operatorname{Re}(\alpha) < 1$ and f in Tate's class of test functions $\pi = \pi^*|\cdot|^{\alpha}$, the <u>local zeta function</u> is defined as follows:

$$\zeta(f,\pi) = \int_K f(x)\pi(x)|x|^{-1}dx .$$

If $f \in \mathcal{S}$ we see that $\zeta(f,\pi) = (f,\pi|\cdot|^{-1})$ so $\zeta(\cdot,\pi) \in \mathcal{S}'$. Defining $\hat{\pi} = \pi^{-1}|\cdot|$ and then $\hat{f}$ essentially as we have defined the

Fourier transform of f, Tate establishes the local zeta function equation:

$$\zeta(f,\pi)\zeta(\hat{g},\hat{\pi}) = \zeta(\hat{f},\hat{\pi})\zeta(g,\pi) \quad .$$

For $f,g \in \mathscr{S}$ this follows from (5.7) since each side is simply $(\pi(-1)/\Gamma(\pi))(f,\pi|\cdot|^{-1})(g,\pi|\cdot|^{-1})$.

Tate defines the factor $\rho(\pi)$ by: $\rho(\pi)\zeta(f,\pi) = \zeta(\hat{f},\hat{\pi})$. (5.7) shows that $\rho(\pi) = \pi(-1)/\Gamma(\pi)$. In this sense, (5.5) is contained in Tates work. (See Lang [1,VII §4].)

Tate's approach suggests an approach to (5.7) by means of a homogeneity argument. The appropriate development is contained in (5.8) - (5.10) together with a proof.

Note that it follows from (5.10) that the only homogeneous distributions are constant multiples of multiplicative characters (in the sense of (5.6)) or are constant multiples of the delta measure.

In (5.10) basic facts about the Dirichlet and conjugate Dirichlet kernel in the theory of Fourier series are used. These results are covered in Zygmund [1,II §5-§8].

In the paragraph preceeding (5.28) we give an overview of Theorems (5.28), (5.29) and (5.31). The characterizations in (5.28) and (5.29) of $J_\pi(u,v)$ as a Mellin transform and Fourier transform, respectively, are formally obvious. The conclusion of (5.31) is clarified if we note that the integral defining the Bessel function is a convolution in the multiplicative structure of the field.

§6. In a certain sense the discussion concerning the ordering of the characters and an identification of "natural orderings" is the crux of the theory of Fourier series; otherwise we would not be able to get far beyond an L^2-theory.

In the study of Walsh-Paley series, one notes that if we construct the Walsh-Paley functions from the Rademacher functions as in I §1, then the 2^nth partial sums are simply averages over dyadic intervals of length 2^{-n}. This fact (which appears as a miracle and/or accident in that setting) indicated that the ordering was a "natural" one. What is shown in §6 is that the Walsh-Paley group (2^ω) has a ring structure that the ring embeds naturally in a field (the 2-series field) and that the characters on 2^ω are in 1:1 correspondence with the (additive) cose of 2^ω in the 2-series field. The ordering we choose and its propertie are consequences of these algebraic facts and the result about 2^nth partial sums for 2^ω is now a special case of (6.7). See also the remar following (6.15) where we show that on p-adic and p-series fields the characters on the ring of integers can be developed by "multiplying out the appropriate "Rademacher" functions in exactly the same way that characters can be developed for the group 2^ω.

It is a fact that 2^nth partial sums (more generally, lacunary partial sums) behave much better than the full sequence of partial sum even for the case of Fourier series on the circle. See Zygmund [1, XIII (1.17) and XV (4.4)] for details.

If we are given any "natural ordering" of the characters, as in (6.3), then we obtain another ordering, just as natural by a rotation. That is, if $\{u(n)\}_{n=0}^{\infty}$ is an ordering of the coset representatives, as in (6.3), then for each $y \in \mathcal{L}^*$, $\{yu(n)\}_{n=0}^{\infty}$ will work just as well, since it is the ordering that would have obtained if we used $\chi' \equiv \chi_y$ as our initial character.

In the remarks following (6.4) and (6.5) we see the same result (namely, Plancherel) appears from two differing points of view. This duplicity was intended as an illumination.

In the example following (6.17) we obtain the result by a trivial application of (6.16) (the "Poisson summation formula") to the fact that $(|x|^{\alpha-1})^\wedge = \Gamma_1(\alpha)|x|^{-\alpha}$ (which is (5.7)). That result was first established by Harper [1] in his study of Hausdorff capacity. Classically, $2^{(\omega)}$ is given the product metric and the integers index index the characters and are given the usual "real" absolute values. In that setting, Harper's result is a computational masterpiece.

The nation of bounded fluctuation, discussed in the text just before (6.19), was introduced by Study [1] in 1896 as a possible substitute for the notion of bounded variation (with which it is equivalent on $\mathbb{R}$).

I used it in my announcement (Taibleson [2], 1967) and later it was used independently by Onneweer [1] and Waterman [1] in 1972 in their study of Vilenkin groups, using approximately the same definitions used in these notes and arriving at similar conclusions. The ring of integers in a local field is one of the simplest non-trivial examples of a Vilenkin group.

With reference to the conclusions of (6.26) on the Cesaro summability of Fourier series, we note that a quite delicate argument shows that $\sigma_n f \to f$ a.e. if $f \in L^1$ and K is of finite characteristic. It is an open question as to whether or not the conclusion holds for K of characteristic zero. The result, for K a p-series field, was announded in Taibleson [2].

(6.29) was established by Fine [2] for the Walsh-Paley group.

The conclusions of (6.34) and (6.35) for Vilenkin groups, are contained in the work of Onneweer [1] and Waterman [1].

* * * * *

P. J. Sally, Jr. has pointed out that all linear functionals on $\mathcal{S}$ are continuous and thus are distributions. This follows easily from the observation that the subspaces of $\mathcal{S}$ that are constant on cosets of $\mathfrak{P}^n$ and are supported on $\mathfrak{P}^{-n}$ $(n > 0)$ are finite dimensional.

Chapter III. Fourier analysis on K^n.

In §1 - §3 of this chapter we extend the fundamental results of Fourier analysis from the one-dimensional to the n-dimensional case, covering the L^1 and L^2 theory in §1 - §2 and the theory of distributions in §3. In §4 - §9 we treat the theory of fractional integration, Riesz and Bessel in some detail. In §10 we take a brief look at the theory of operators that commute with translations.

1. The L^1 theory

We will record here the main facts needed for analysis on K^n, the n-dimensional vector spaces over K. For the most part, the material in §1 - §3 of this chapter is a transplantation of §1 - §3 of Chapter II. The major exception is the material on multipliers and convolutions in §3 of this chapter. Most proofs are omitted, since they would be trivial reworkings of the proofs for $n = 1$.

Let $x = (x_1, x_2, \ldots, x_n) \in K^n$; $x_i \in K$, $i = 1,2,\ldots,n$.

Let $|x| = \sup_i |x_i|$. It is easy to see that $x \to |x|$ is an ultrametric valuation on K^n. That is, $x \to |x|$ is a norm and $|x+y| \leq \max[|x|,|y|]$.

Haar measure on K^n is given by $dx = dx_1 \cdots dx_n$.

By an abuse of notation we denote the balls of radius q^{-k} by $\mathfrak{P}^k$, $k \in \mathbb{Z}$, letting $\mathfrak{P} = \mathfrak{P}^1$ and $\mathfrak{O} = \mathfrak{P}^0$. Thus, $\mathfrak{P}^k = \{x \in K^n: |x| \le q^{-k}\}$. We also let $\mathfrak{O}^* = \{x \in K^n: |x| = 1\}$.

The measure (Borel-Lebesgue) of a measurable set E in K^n is denoted $|E|$. It is easy to see that $|\mathfrak{P}^k| = q^{-kn}$, $|\mathfrak{O}| = 1$, $|\mathfrak{O}^*| = 1-q^{-n}$, $|\mathfrak{P}| = q^{-n}$. It is also easy to see that if $\alpha \in K$ then $d(\alpha x) = |\alpha|^n dx$, so that $|x|^{-n}dx$ is invariant under scalar multiplication by $\alpha \in K^*$.

The following facts restate some of the material at the end of Chapter I:

For $x, u \in K^n$ let $x \cdot u = x_1 u_1 + \cdots + x_n u_n$. Let χ be the character on K that was fixed in Chapter II. Then we write $\chi_u(x) = \chi(u \cdot x) = \chi(u_1 x_1 + \cdots + u_n x_n) = \prod_{k=1}^n \chi_{u_k}(x_k)$. For each $u \in K^n$, χ_u is a character on K^n; that is, $\chi_u \in \hat{K}^n$. With vector addition in $\hat{K}^n$ defined by $\chi_u \oplus \chi_v = \chi_{u+v}$ and scalar multiplication by $\alpha \chi_u = \chi_{\alpha u}$ we have that $u \to \chi_u$ is topological isomorphsim of K^n and $\hat{K}^n$ as vector spaces. Thus K^n is self dual.

We now add several obvious facts:

If $u \in \mathfrak{O}^*$ then χ_u is trivial on $\mathfrak{O}$ but it non-trivial on $\mathfrak{P}^{-1}$. $\mathfrak{P}^k$ is a subgroup of K^n. If we denote $\mathfrak{P}^k$ in K^n by $\mathfrak{P}^k_{(n)}$ then $\mathfrak{P}^k_{(n)} = \mathfrak{P}^k_{(1)} \times \cdots \times \mathfrak{P}^k_{(1)}$. Hence if $\Phi_k = \Phi_k^{(n)}$ is the characterist function of $\mathfrak{P}^k = \mathfrak{P}^k_{(n)}$ we see that $\Phi^{(n)}_k(x) = \Phi^{(1)}_k(x_1)\cdots\Phi^{(1)}_k(x_n)$. That is, "the balls in K^n are cubes".

The L^p spaces, $1 \le p \le \infty$, and the space of continuous functions, C_0 are defined and normed in the usual way.

Examples. Let $f_\alpha(x) = |x|^\alpha \Phi_0(x)$, $g_\alpha = |x|^\alpha (1 - \Phi_0(x))$. Then $f_\alpha \in L^p$, $1 \le p < \infty$, iff $\alpha + n/p > 0$. $g_\alpha \in L^p$, $1 \le p < \infty$, iff $\alpha + n/p < 0$.

The Fourier transform for $f \in L^1(K^n)$ is defined in the usual way: $\hat{f}(x) = \int_{K^n} f(\xi)\overline{\chi}_x(\xi)d\xi$.

Theorem (1.1). (a) $f \to \hat{f}$ is a bounded linear map from L^1 to L^∞, $\|\hat{f}\|_\infty \le \|f\|_1$. (b) If $f \in L^1$. $\hat{f}$ is uniformly continuous.

If we let $(\tau_h f)(x) = f(x-h)$, $h \in K^n$,

$$(1.2) \qquad (\tau_h f)^\wedge = \overline{\chi}_h \hat{f}, \quad (\chi_h f)^\wedge = \tau_h \hat{f}, \quad f \in L^1(K^n).$$

Sets of the form $h + \mathfrak{P}^k$ are spheres with center h and radius q^{-k}. The measure of the sphere is q^{-kn}, its characteristic function is $\tau_h \Phi_k$.

We define $\mathcal{S} = \mathcal{S}(K^n)$ to be the space of finite linear combinations (with complex coefficients) of characteristic functions of spheres.

(1.3) $\tau_h\Phi_k$ <u>is a continuous function with compact support.</u>

$\mathcal{S}$ <u>is dense in</u> L^p, $1 \leq p < \infty$ <u>and in</u> C_0.

(1.4) <u>For all</u> $k \in \mathbb{Z}$, $\hat{\Phi}_k = q^{-kn}\Phi_{-k}$.

<u>Proof</u>. $\Phi_k^{(n)}(x) = \Phi_k^{(1)}(x_1)\cdots\Phi_k^{(1)}(x_n)$, so

$$\hat{\Phi}_k^{(n)}(\xi) = \hat{\Phi}_k^{(1)}(\xi_1)\cdots\hat{\Phi}_k^{(1)}(\xi_n) = (q^{-k}\Phi_{-k}^{(1)}(\xi_1))\cdots\cdot(q^{-k}\Phi_{-k}^{(1)}(\xi_n))$$

$$= q^{-kn}\Phi_{-k}^{(n)}(\xi) .$$

For $k \in \mathbb{Z}$ define $\delta_k f$ by $\delta_k f(x) = f(\mathfrak{p}^k x)$ $(\mathfrak{p} \in K)$.

(1.5) $(\delta_k f)\hat{} = q^{kn}(\delta_{-k}\hat{f})$.

(1.6) (Riemann-Lebesgue) <u>If</u> $f \in L^1(K^n)$ <u>then</u> $\hat{f}(x) \to 0$ <u>as</u> $|x|$

As for $n = 1$, the Fourier transform extends to finite Borel measures, (1.1) holds but (1.6) fails.

Convolution is defined in the usual way and we have:

(1.7) <u>If</u> $f \in L^p(K^n)$, $1 \leq p \leq \infty$, $g \in L^1(K^n)$ <u>then</u> $f * g$ <u>is defined a.</u>

$f * g \in L^p$, $\|f * g\|_p \leq \|f\|_p\|g\|_1$.

Similarly, if $f \in L^p$, $1 \leq p < \infty$; or $f \in C_0$, $p = \infty$, μ is a finite Borel measure then $f * \mu \in L^p$, $1 \leq p < \infty$; $f * \mu \in C_0$, $p = \infty$ and $\|f * \mu\|_p \leq \|f\|_p\|\mu\|_M$ where $\|\mu\|_M$ is the total variation of μ.

By the usual argument we obtain:

(1.8) $f, g \in L^1 \Longrightarrow (f * g)^\wedge = \hat{f}\hat{g}$.

$f \in L^1$, μ *a finite Borel measure* $\Longrightarrow (f * \mu)^\wedge = \hat{f}\hat{\mu}$.

We turn to the problem of inversion of Fourier transforms. For g locally integrable on K^n let

$$A_k(g) = \int_{K^n} g\,\Phi_{-k} = \int_{|z| \le q^k} g(x), \quad k \in \mathbb{Z}.$$

(1.9) (Multiplication) $\int \hat{f} g = \int f \hat{g}$.

(1.10) *If* $f \in L^1$, $k \in \mathbb{Z}$ *then*

$$A_k(\hat{f}\,\chi_x) = q^{kn} \int_{|x-z| \le q^{-k}} f(z)\,dz = (f * q^{kn}\Phi_k)(x) .$$

(1.11) *If* S *and* T *are spheres either they are disjoint or one contains the other.*

(1.12) *Let* F *be a measurable subset of* K^n *of finite measure.* $\mathfrak{J} = \{T_\alpha\}$ *be a cover of* F *by spheres. Given* λ, $0 < \lambda < 1$, *there is a finite sub-collection* $\{T_k\}_{k=1}^N$ *of mutually disjoint spheres such that* $\sum_{k=1}^N |T_k| > \lambda |F|$.

The *Hardy-Littlewood maximal function* is defined as before: If f is locally integrable,

$$Mf(x) = \sup_{\substack{S,\text{spheres}\\ x\in S}} \frac{1}{|S|}\int_S |f| = \sup_{k\in\mathbb{Z}} q^{kn}\int_{|x-z|<q^{-k}} |f(z)|\,dz\ .$$

(1.13) <u>If</u> $f \in L^1$, $s > 0$ <u>then</u> $|\{x: Mf(x) > s\}| \leq s^{-1}\|f\|_1$.

If f is locally integrable, $x \in K^n$ is a <u>regular point of</u> f if

$$q^{kn}\int_{|x-z|\leq q^{-k}} f(z)\,dz \to f(x) \text{ as } k \to \infty\ .$$

(1.14) <u>If</u> f <u>is locally integrable then for a.e.</u> x, x <u>is a regular point of</u> f.

(1.15) <u>If</u> $f \in L^1$ <u>then</u> $A_k(\chi_x\hat{f}) \to f(x)$ <u>at each point of regularity of</u> f, <u>and hence a.e. and in particular at each point of continuity of</u> f.

(1.16) <u>If</u> $f, \hat{f} \in L^1$ <u>then</u> f <u>is equal a.e. to the continuous function,</u> $g(x) = \int_{K^n} \hat{f}(\xi)\chi_x(\xi)\,d\xi$.

(1.17) <u>If</u> $f,g \in L^1$ <u>and</u> $\hat{f} = \hat{g}$ <u>then</u> $f(x) = g(x)$ a.e.

(1.18) <u>If</u> $f \in L^1$, $\hat{f} > 0$, f <u>is continuous at</u> 0 <u>then</u> $\hat{f} \in L^1$ <u>and</u>

$$\int \hat{f} = f(0).$$

Let $R(x,k) = q^{-kn}\,\Phi_{-k}$, $x \in K^n$, $k \in \mathbb{Z}$.

(1.19) $\hat{R}(\cdot,k) = \Phi_k \, , \, \hat{\Phi}_k = R(\cdot,k)$

(1.20) $R(\cdot,k) * R(\cdot,\ell) = R(\cdot,k \vee \ell)$

2. The L^2 theory

(2.1) $f \in L^1 \cap L^2 \Longrightarrow \|\hat{f}\|_2 = \|f\|.$

For $f \in L^2$, $\hat{f} \in L^2$ is then defined as the limit in the L^2-norm of $(f\Phi_{-k})^\wedge$ as $k \to \infty$.

(2.2) (Multiplication) $f,g \in L^2 \Longrightarrow \int f\hat{g} = \int \hat{f} g$.

(2.3) <u>The Fourier transform is unitary on</u> $L^2(K^n)$.

The inverse Fourier transform is denoted $f \to f^\vee$. Let $\tilde{f}(x) = f(-x)$.

(2.4) $f \in L^2 \Longrightarrow f^\vee = (\tilde{f})^\wedge$.

(2.5) $f,g \in L^2 \Longrightarrow \int f\,\bar{g} = \int \hat{f}\,\bar{\hat{g}}$

(2.6) $f \in L^2 \Longrightarrow \lim_{k \to \infty} \int_{|\xi| \le q^k} f(\xi)\bar{\chi}_x(\xi)d\xi = \hat{f}(x)$ a.e.

The Fourier transform is then extended to L^p, $1 \le p \le 2$.

(2.7) $f \in L^1;\ g \in L^p,\ 1 \leq p \leq 2 \implies (f * g)^\wedge = \hat{f}\,\hat{g}$ a.e. .

3. Distributions on K^n

Let $\mathcal{S}$ be defined as in §1 .

(3.1) $\varphi \in \mathcal{S}$ *iff there are integers* k *and* ℓ *such that* φ *is constant on cosets of* $\mathfrak{P}^k$ *and is supported on* $\mathfrak{P}^\ell$.

(3.2) $\varphi \in \mathcal{S}$ *is constant on cosets of* $\mathfrak{P}^k$ *and supported on* $\mathfrak{P}^\ell$ *iff* $\hat{\varphi}$ *is constant on cosets of* $\mathfrak{P}^{-\ell}$ *and is supported on* $\mathfrak{P}^{-k}$.

(3.3) *If* $\varphi, \psi \in \mathcal{S}$ *then* $\tau_h\varphi$, $\tilde{\varphi}$ *and* $\varphi * \psi \in \mathcal{S}$.

$\mathcal{S}$ is given a topology as a topological vector space, as before.

(3.4), (3.5) $\mathcal{S}$ *is complete and separable*.

(3.6), (3.7) $\varphi \to \hat{\varphi}$; $\varphi \to \tau_h\varphi$, $h \in K^n$, $\varphi \to \tilde{\varphi}$ *are homeomorphisms of* $\mathcal{S}$ *onto* $\mathcal{S}$.

(3.8) $\mathcal{S}$ *is continuously contained and dense in* L^p, $1 \leq p < \infty$ *and in* C_0 .

$\mathcal{S}'(K^n)$ is the space of distributions on K^n. It is the topological dual of $\mathcal{S}$ with the Weak* topology. The action of $f \in$

on $\varphi \in \mathcal{S}$ is denoted (f,φ). The use of the locutions: $f \in \mathcal{S}'$ is a: function, measure etc, will be maintained.

We define the Fourier transform and inverse Fourier transforms for $f \in \mathcal{S}'$ by the rules: $(\hat{f},\varphi) = (f,\hat{\varphi})$, $(f^{\vee},\varphi) = (f,\varphi^{\vee})$ for all $\varphi \in \mathcal{S}$. Clearly,

(3.9) $(\hat{f})^{\vee} = f = (f^{\vee})^{\wedge}$ <u>for all</u> $f \in \mathcal{S}'$.

For $h \in K^n$ we define translates of f by h, $\tau_h f$ by $(\tau_h f,\varphi) = (f,\tau_{-h}\varphi)$ and reflection $\tilde{f}$ by $(\tilde{f},\varphi) = (f,\tilde{\varphi})$.

(3.10),(3.11) $f \to \hat{f}$, $f \to f^{\vee}$, $f \to \tau_h f$, $f \to \tilde{f}$ <u>are homeomorphisms of</u> $\mathcal{S}'$ <u>onto</u> $\mathcal{S}'$.

<u>Definition</u>. For $f \in \mathcal{S}'$, $\varphi \in \mathcal{S}$ we define $f\varphi = \varphi f$ by $(f\varphi, \psi) = (f,\varphi\psi)$ for all $\psi \in \mathcal{S}$.

<u>Proposition (3.12)</u>. <u>For</u> $\psi,\varphi \in \mathcal{S}$, $f \in \mathcal{S}'$, <u>the maps</u> $f \to f\varphi$, $\psi \to \psi\varphi$ <u>are continuous maps from</u> $\mathcal{S}'$ <u>into</u> $\mathcal{S}'$ <u>and</u> $\mathcal{S}$ <u>into</u> $\mathcal{S}$ <u>respectively</u>.

<u>Proof</u>. The proof for $\psi \to \psi\varphi$ is simple. For $f \to f\varphi$ is follows from that case and the definition.

<u>Definition</u>. $\mathcal{O}_M$ is the class of <u>locally constant functions</u>. $f \in \mathcal{O}_M$ if f is a function that is constant on the cosets of some $\mathfrak{P}^k$.

<u>Proposition (3.13)</u>. $f \in \mathcal{O}_M$ <u>iff</u> $f \in \mathcal{S}'$ <u>and there is a</u> $k \in \mathbb{Z}$ <u>such that</u> $\tau_x f = f$ <u>for all</u> $x \in \mathfrak{P}^k$.

<u>Proof</u>. If $f \in \mathcal{O}_M$ and is constant on cosets of $\mathfrak{P}^k$ then for all $x \in \mathfrak{P}^k$, $\tau_x f = f$ as a function and so as an element of $\mathcal{S}'$.

Now suppose $f \in \mathcal{S}'$ and $\tau_x f = f$ for all $x \in \mathfrak{P}^k$. Let $g(y) = q^{kn}(f,\tau_y \Phi_k)$. $g \in \mathcal{O}_M$ since if $x \in \mathfrak{P}^k$,

$$(\tau_x g)(y) = q^{kn}(f,\tau_{y-x}\Phi_k) = q^{kn}(\tau_x f,\tau_y \Phi_k) = q^{kn}(f,\tau_y \Phi_k) = g(y).$$

We also see that f and g agree, as elements of $\mathcal{S}'$, on any function of the form $\tau_h \Phi_m$, $m \leq k$. Also, if $\ell \geq 0$,

$(g,\tau_h \Phi_{k+\ell}) = q^{-(k+\ell)n} g(h) = q^{-\ell n}(f,\tau_h \Phi_k)$. We will show that $(f,\tau_h \Phi_{k+\ell}) = q^{-\ell n}(f,\tau_h \Phi_k)$.

There are $q^{\ell n}$ coset representatives of $\mathfrak{P}^{k+\ell}$ in $\mathfrak{P}^k$, $\{a_\nu\}$.

$$(f,\tau_h \Phi_k) = \sum_\nu (f,\tau_{h-a_\nu}\Phi_{k+\ell}) = \sum_\nu (\tau_{a_\nu} f, \tau_h \Phi_{k+\ell})$$

$$= \sum_\nu (f,\tau_h \Phi_{k+\ell}) = q^{\ell n}(f,\tau_h \Phi_{k+\ell}) .$$

<u>Definition</u>. For $f \in \mathcal{S}'$, $\varphi \in \mathcal{S}$, $f * \varphi \in \mathcal{S}'$ is defined by $(f * \varphi)^\wedge = \hat{f}\hat{\varphi}$

<u>Proposition (3.14)</u>. <u>For</u> $f \in \mathcal{S}'$, $\varphi \in \mathcal{S}$, <u>the map</u> $f \to f * \varphi$ <u>is a continuous map from</u> $\mathcal{S}'$ <u>into</u> $\mathcal{S}'$.

Proof. Immediate from (3.12).

Fix $\varphi,\psi \in \mathcal{S}$, $f \in \mathcal{S}'$, $x \in K^n$. Then the following facts are easily established and serve as motivation for (3.15).

$$(\tau_{-x}\varphi)^{\sim} = \tau_x\tilde{\varphi}\ ,\ (\tau_{-x}f)^{\sim} = \tau_x\tilde{f}\ ,\ (\varphi * \psi)^{\sim} = \tilde{\varphi} * \tilde{\psi},\ (\tilde{\varphi} * \psi)^{\sim} = \varphi * \tilde{\psi}\ .$$

Theorem (3.15). The following three characterizations of $f * \varphi \in \mathcal{S}'$ for $f \in \mathcal{S}'$, $\varphi \in \mathcal{S}$ are equivalent:

(i) $(f * \varphi)^{\wedge} = \hat{f}\,\hat{\varphi}$

(ii) $(f * \varphi, \psi) = (f, \tilde{\varphi} * \psi)$

(iii) $(f * \varphi)(x)$ is that function $g \in \mathcal{O}_M$ defined by

$$g(x) = (f, \tau_x\tilde{\varphi}) = (\tau_x\tilde{f}, \varphi)\ .$$

Proof. (ii) $\Longrightarrow$ (i). $((f * \varphi)^{\wedge}, \psi) = (f * \varphi, \hat{\psi}) = (f, \tilde{\varphi} * \psi)$

$= (\hat{f}, (\tilde{\varphi} * \hat{\psi})^{\vee}) = (\hat{f}, (\tilde{\varphi})^{\vee}(\hat{\psi})^{\vee}) = (\hat{f}, \hat{\varphi}\,\psi) = (\hat{f}\,\hat{\varphi}, \psi)$.

(i) $\Longrightarrow$ (ii) $((f * \varphi), \psi) = ((f * \varphi)^{\wedge}, \psi^{\vee}) = (\hat{f}\,\hat{\varphi}, \psi^{\vee})$

$= (\hat{f}, \hat{\varphi}\,\psi^{\vee}) = (f, (\hat{\varphi} * \psi)^{\vee}) = (f, \tilde{\varphi} * \psi)$.

(ii) $\Longleftrightarrow$ (iii) and $(f, \tau_x\tilde{\varphi}) \in \mathcal{O}_M$.

We wish to show that $(f, \tilde{\varphi} * \psi) = ((f, \tau_{\cdot}\tilde{\varphi}), \psi)$ for all $\varphi, \psi \in \mathcal{S}$. From the linearity of both forms in φ and ψ it will suffice to assume that $\varphi = \tau_u\Phi_k$ and $\psi = \tau_v\Phi_k$ for some $k \in \mathbb{Z}$, $u,v \in K^n$. Note that $\tilde{\varphi} = \tau_{-u}\Phi_k$, $\tilde{\psi} = \tau_{-v}\Phi_k$, and so $(f, \tilde{\varphi} * \psi) = q^{-kn}(f, \tau_{v-u}\Phi_k)$. Now note that if φ is constant on cosets of $\mathfrak{P}^k$, and $y \in \mathfrak{P}^k$ then $g(x+y) = (f,\tau_{x+y}\tilde{\varphi}) = (f,\tau_x\tilde{\varphi}) = g(x)$, so $g \in \mathcal{O}_M$. Now let $g(x) = (f,\tau_x(\tau_u\Phi_k)^{\sim}) = (f,\tau_{x-u}\Phi_k)$, so the value of g on the set $v + \mathfrak{P}^k$ is $g(v) = (f,\tau_{v-u}\Phi_k)$ and hence, $(g,\tau_v\Phi_k) = q^{-kn}(f,\tau_{v-u}\Phi_k)$.

In defining the product $f\varphi, f \in \mathscr{S}', \varphi \in \mathscr{S}$ the important point was that $\varphi\psi \in \mathscr{S}$ for all $\psi \in \mathscr{S}$. But this also holds for all $\varphi \in \mathcal{O}_M$. In fact, if we define $f\varphi = \varphi f$, $f \in \mathscr{S}'$, $\varphi \in \mathcal{O}_M$ by $(f\varphi, \psi) = (f,\varphi\psi)$ for all $\psi \in \mathscr{S}$, then ,

<u>Proposition (3.16)</u>. <u>The map</u> $f \rightarrow f\varphi, \varphi \in \mathcal{O}_M$ <u>is a continuous map of</u> $\mathscr{S}'$ <u>into</u> $\mathscr{S}'$.

<u>Proof</u>. This follows from the observation that $\psi \rightarrow \psi\varphi$ is continuous map of $\mathscr{S}$ into $\mathscr{S}$.

<u>Definition.</u> Let $\mathcal{O}_C$ be the space of distributions with compact support. That is, $f \in \mathcal{O}_C$ iff $f \in \mathscr{S}'$ and there is a $k \in \mathbb{Z}$ such that $f\Phi_k = f$.

<u>Proposition (3.17)</u>. $f \in \mathcal{O}_M$ <u>iff</u> $\hat{f} \in \mathcal{O}_C$. <u>In particular if</u> $f \in \mathscr{S}'$ <u>then</u> $\tau_x f = f$ <u>for all</u> $x \in \mathfrak{P}^k$ <u>iff</u> $\Phi_{-k}\hat{f} = \hat{f}$.

<u>Proof</u>. Suppose $f \in \mathcal{O}_M$ and $\tau_x f = f$ for all $x \in \mathfrak{P}^k$. Then f is a function that is constant on cosets of $\mathfrak{P}^k$. Let $\varphi \in \mathscr{S}$.

Then $(\Phi_{-k}\hat{f})^{\vee}(x) = q^{kn}(\Phi_k * f)(x) = q^{kn}(f,\tau_x\Phi_k) = q^{kn}q^{-kn}f(x) = f(x)$.

Thus $\Phi_{-k}\hat{f} = \hat{f}$.

Now suppose $\hat{f} \in \mathcal{O}_C$ and $\hat{f}\Phi_{-k} = \hat{f}$. Then for all $\Phi \in \mathcal{S}$, $x \in \mathfrak{B}^k$, $(\tau_x f,\varphi) = (f,\tau_{-x}\varphi) = (\hat{f},(\tau_{-x}\varphi)^{\vee}) = (\hat{f},\chi_x\varphi^{\vee})$
$= (\hat{f}\Phi_{-k},\chi_{-x}\varphi^{\vee}) = (\hat{f},\Phi_{-k}\chi_{-x}\varphi^{\vee}) = (\hat{f}\Phi_{-k},\varphi^{\vee}) = (\hat{f},\varphi^{\vee}) = (f,\varphi)$.

Remark. (3.17) is an extension of (3.2) to $\mathcal{S}'$.

Definition. If $f \in \mathcal{S}'$, $\varphi \in \mathcal{O}_C$ then $f*\varphi \in \mathcal{S}'$ is defined by $(f*\varphi)^{\wedge} = \hat{f}\hat{\varphi}$.

Proposition (3.18). *If $\varphi \in \mathcal{O}_C$ the map $f \to f*\varphi$ is a continuous map from $\mathcal{S}'$ into $\mathcal{S}'$.*

Proof. An immediate consequence of (3.16) and (3.17).

The next result follows from a gathering earlier results.

Theorem (3.19). (a) *Let $f_1,f_2,\ldots,f_s \in \mathcal{S}'$, all but (at most) one being in $\mathcal{O}_M$. Then $f_1f_2\cdots f_s$ is well defined as a commutative and associative product.*

(b) *Let $f_1,\ldots,f_s \in \mathcal{S}'$ all but (at most) one being in $\mathcal{O}_C$. Then $f_1*\cdots*f_s$ is well defined as a commutative and associative convolution product.*

4. Riesz fractional integration

Definition. $\Gamma_n(\alpha) = (1 - q^{\alpha-n})/(1 - q^{-\alpha}), \ \alpha \in C, \ \alpha \neq 0$.

Remark. $\Gamma_n(\alpha)\Gamma_n(n-\alpha) = 1, \ \alpha \neq 0, n$.

Remark. $\Gamma_n(\alpha)$ is meromorphic with a unique simple zero at $\alpha = n$ and unique simple pole at $\alpha = 0$. Its residue at $\alpha = 0$ is $(1-q^{-n})/\ell n$ at $\alpha = 0$. $\Gamma_n(\alpha)$, for $n = 1$, is the function Γ_1 of II.§5 .

Lemma (4.1).

$$\int_{|x| = q^k} \chi(u \cdot x)dx = \begin{cases} q^{kn}(1-q^{-n}), & |u| \leq q^{-k} \\ -q^{kn}q^{-n}, & |u| = q^{-k+1} \\ 0, & |u| > q^{-k+1} \end{cases}$$

Proof. The usual argument will work.

Theorem (4.2). *If* $\operatorname{Re}(\alpha) > 0$ *and* $|u| = 1$ *then*

$$\Gamma_n(\alpha) = P\int |x|^{\alpha-n}\, \bar{\chi}(u \cdot x)dx$$

$$= \lim_{k \to \infty} \int_{q^{-k} \leq |x| \leq q^k} |x|^{\alpha-n}\, \bar{\chi}(u \cdot x)dx \ .$$

Proof. Using (4.1) we see that the integral is equal to

$$\sum_{k=0}^{\infty} q^{-k(\alpha-n)}q^{-kn}(1-q^{-n}) - q^{\alpha-n}$$

$$= (1-q^{-n})\sum_{k=0}^{\infty} q^{-k\alpha} - q^{\alpha-n} = (1-q^{-n})/(1-q^{-\alpha}) - q^{\alpha-n}$$

$$= (1-q^{\alpha-n})/(1-q^{-\alpha}).$$

The function $|x|^{\alpha-n}$ determines a distribution as follows:

If $\operatorname{Re}(\alpha) > 0$ then $|x|^{\alpha-n}$ is locally integrable so $(|x|^{\alpha-n},\varphi) = \int |x|^{\alpha-n}\varphi(x)dx$, $\varphi \in \mathscr{S}$. If $\varphi \in \mathscr{S}$ and φ is constant on cosets of $\mathfrak{B}^k$ and is supported on $\mathfrak{B}^\ell$ (we may assume $\ell \leq k$), and $\operatorname{Re}(\alpha) > 0$ then

$$f(\alpha) = (|x|^{\alpha-n},\varphi) = \varphi(0)((1-q^{-n})/(1-q^{-\alpha}))q^{-k\alpha} + \int_{q^{-k}<|x|\leq q^{-\ell}} \varphi(x)|x|^{\alpha-n}dx \quad , \tag{4.3}$$

is analytic and can be extended as a meromorphic function. If $\varphi(0) = 0$ it is entire. If $\varphi(0) \neq 0$ then it has a simple pole at $\alpha = 0$ with residue $\varphi(0)(1-q^{-n})/\ell n\, q$.

If we define $(|x|^{\alpha-n},\varphi)$ its extension of (4.3) (except possibly at $\alpha = 0$) we see that $|x|^{\alpha-n} \in \mathscr{S}'$ if $\alpha \neq 0$.

Taking advantage of the behaviour of $\Gamma_n(\alpha)$ near $\alpha = 0$ we see that $(1/\Gamma_n(\alpha)|x|^{\alpha-n},\varphi)$ is analytic in a neighborhood of $\alpha = 0$ and has a removable singularity there. Its value at $\alpha = 0$ is $\varphi(0)$. Hence $(1/\Gamma_n(\alpha)|x|^{\alpha-n},\varphi)|_{\alpha=0} = (\delta,\varphi)$. We adopt the convention $(1/\Gamma_n(0))|x|^{-n} = \delta$. With this convention we have,

__Lemma (4.4)__. $|x|^{\alpha-n} \in \mathscr{S}'$, $\alpha = 0$; $(1/\Gamma_n(\alpha))|x|^{\alpha-n} \in \mathscr{S}'$, $\alpha \neq n$.

Theorem (4.5). As elements of $\mathscr{S}'$, $((1/\Gamma_n(\alpha))|x|^{\alpha-n})^\wedge = |x|^{-\alpha}$, $\alpha \neq n$.

Proof. Since $\hat{\delta} = 1$ the case $\alpha = 0$ is taken care of by our convention. Thus, we may assume that $\alpha \neq 0, n$.

We need to show that $((1/\Gamma_n(\alpha))|x|^{\alpha-n},\hat{\varphi}) = (|x|^{-\alpha},\varphi)$ for all $\varphi \in \mathscr{S}$. Since both sides are analytic in α ($\alpha \neq 0,n$) for φ fixed, it will suffice to show that the result holds when $0 < \alpha < n$. As in II(5.7) we can establish this result by a direct calculation.

Suppose $\alpha, \beta, \alpha + \beta \neq 0, n$. Then we see that as elements of $\mathscr{S}'$

$$((1/\Gamma_n(\alpha))|x|^{\alpha-n})^\wedge((1/\Gamma_n(\beta))|x|^{\beta-n})^\wedge = |x|^{-(\alpha+\beta)}$$
$$= ((1/\Gamma_n(\alpha+\beta))|x|^{\alpha+\beta-n})^\wedge .$$

One is thus led to the formal relation

$$(4.6) \qquad |x|^{\alpha-n} * |x|^{\beta-n} = \left[(\Gamma_n(\alpha)\Gamma_n(\beta))/\Gamma_n(\alpha+\beta)\right] |x|^{\alpha+\beta-n} .$$

This leads to consideration of the integral,

$$(4.7) \qquad \int |u-x|^{\alpha-n}\, |x|^{\beta-n}\, dx , \quad u \neq 0 .$$

If $0 < \mathrm{Re}(\alpha), \mathrm{Re}(\beta), \mathrm{Re}(\alpha+\beta) < n$ it is easy to see that (4.7) converges absolutely. Moreover, we see that it is equal to

$$|u|^{\alpha+\beta-n} \int |u'-x|^{\alpha-n}\, |x|^{\beta-n}\, dx, \quad u' = u|u|^{-1}, \; u' \in \Theta^*.$$

Definition. $B_n(\alpha,\beta) = \Gamma_n(\alpha)\Gamma_n(\beta)/\Gamma_n(\alpha+\beta)$.

<u>Theorem (4.8)</u>. <u>If</u> $0 < \mathrm{Re}(\alpha), \mathrm{Re}(\beta), \mathrm{Re}(\alpha+\beta) < n$, $u \in \mathfrak{O}^*$, <u>then</u> $B_n(\alpha,\beta) = \int |u-x|^{\alpha-n}\,|x|^{\alpha-n}\,dx$.

<u>Proof.</u> $\int_{K^n} |u-x|^{\alpha-n}|x|^{\beta-n}dx$

$$= \int_{|x|<1} |x|^{\beta-n} + \int_{|x|<1} |x|^{\alpha-n} + \int_{\substack{|x|=1\\|u-x|=1}} 1\,dx + \int_{|x|>1} |x|^{\alpha+\beta-n}\,dx$$

$$= (1-q^{-n})q^{-\beta}/(1-q^{-\beta}) + (1-q^{-n})q^{-\alpha}/(1-q^{-\alpha})$$

$$+ (1-2q^{-n}) + (1-q^{-n})q^{\alpha+\beta-n}/(1-q^{\alpha+\beta-n})$$

$= B_n(\alpha,\beta)$ by a miserable calculation.

We could also get this result by a distribution argument as in II.

Since $|x|^{\alpha-n} \in \mathcal{S}'$, $\alpha \neq 0$, $1/\Gamma_n(\alpha)|x|^{\alpha-n} \in \mathcal{S}'$, $\alpha \neq n$, the map $\varphi \to (1/\Gamma_n(\alpha))|x|^{\alpha-n} * \varphi = I^\alpha\varphi$, $\alpha \neq n$ (so $(I^\alpha\varphi)^\wedge = |x|^{-\alpha}\hat{\varphi}$) is well defined.

If we only knew what we meant when we wrote $I^\beta(I^\alpha\varphi)$, $\varphi \in \mathcal{S}$ we would have $I^\beta I^\alpha = I^{\alpha+\beta}$, and we would have a semi-group of operators. There are two difficulties. The existence problem, and the problem of "n". We deal with existence problem in this section. The problem of "n" will be treated in the next section where we "smooth" the kernels.

We consider the integral

$$I^\alpha f(x) = (1/\Gamma_n(\alpha))\int f(z)|x-z|^{\alpha-n}dz \; .$$

Theorem (4.9). (Soboleff) Suppose $0 < \text{Re}(\alpha) < n$, $1 \leq p < r < \infty$, $0 < 1/r = (1/p) - (\text{Re}(\alpha)/n)$.

(a) If $f \in L^p$, $I^\alpha f(x)$ exists a.e.

(b) If $f \in L^p$, $p > 1$, $I^\alpha f \in L^r$, $\|I^\alpha f\|_r \leq A_{pr}\|f\|_p$, $A_{pr} > 0$ is independent of f.

(c) If $f \in L^1$, $s > 0$, then $|\{x : |I^\alpha f(x)| > s\}| \leq (A_r\|f\|_1\, s^{-1})^r$, $A_r > 0$ independent of f.

Proof. The proof follows classical lines. For antecedents and comments see §11.

Let $K(x) = |x|^{\alpha-n}$. We show that if $f \in L^p$, $f * K(x)$ exists a.e. Let $K_1 = K\Phi_{-k}$, $K_\infty = K(1-\Phi_{-k})$. Then $f * K = f * K_1 + f * K_\infty$. Since K is locally integrable, $K_1 \in L^1$ so $f * K_1(x)$ exists a.e. by (1.7)

Let p' be the index conjugate to p. That is $(1/p') + (1/p) = 1$, so $p' = p/(p-1)$ $(p = 1 \Longrightarrow p' = \infty)$. Note that $K_\infty \in L^{p'}$. (For $p = 1$ note that $\text{Re}(\alpha) - n < 0$. For $p > 1$ we have $\text{Re}((\alpha-n)p') < -n$.) It follows from Hölders inequality that $f * K_\infty(x)$ exists for all x, and so $f * K(x)$ exists a.e.

This takes care of (a). To complete the proof we will show that $f \to I^\alpha f$ is of weak type (p,r). That is, $f \in L^p$, $0 < 1/r = (1/p) - (\text{Re}(\alpha)/n)$ implies that for all $s > 0$, $|\{x : |K * f(x)| > s\}| \leq (A_{pr}\|f\|_p\, s^{-1})^r$, $A_{pr} > 0$.

The special case $p = 1$ is (c). Part (b) follows by the Marcinkiewicz interpolation theorem.

Fix $s > 0$. We may assume $\|f\|_p = 1$

$$|\{x : |K * f(x)| > s\}| \le |\{x : |K_1 * f(x)| > s/2\}| + |\{x : |K_\infty * f(x)| > s/2\}| .$$

$$|K_\infty * f(x)| \le \|K_\infty * f\|_\infty \le \|K_\infty\|_{p'} .$$

For $p = 1$, $\|K_\infty\| \le q^{(k+1)(\mathrm{Re}(\alpha)-n)} < q^{-kn/r}$. If $p > 1$,

$$\|K_\infty\|_{p'} = \Big[\int_{|x| > q^k} |x|^{(\mathrm{Re}(\alpha)-n)p'} dx \Big]^{1/p'}$$

$$= \Big[\sum_{\ell=k+1}^{\infty} q^{\ell(\mathrm{Re}(\alpha)-n)p'} q^{\ell n} (1-q^{-n}) \Big]^{1/p'}$$

$$\le (1-q^{-n})^{1/p'} \Big[\sum_{\ell=k+1}^{\infty} q^{((\mathrm{Re}(\alpha)-n)p'+n)\ell} \Big]^{1/p'}$$

$$= (1-q^{-n})^{1/p'} \Big[\sum_{\ell=k+1}^{\infty} q^{-(np'/r)\ell} \Big]^{1/p'}$$

$$= (1-q^{-n})^{1/p'} q^{-nk/r} q^{-n/r} (1-q^{-np'/r})^{-1/p'} = c_1 q^{-kn/r} .$$

Thus, $|K_\infty * f(x)| \le c_1 q^{-nk/r}$. Choosing

$k = \Big[\frac{r}{n} \log_q \big(\frac{s}{2c_1} \big) \Big] + 1$, we have $|K_\infty * f(x)| < s/2$.

Thus, $|\{x : |K * f(x)| > s\}| \le |\{x : |K_1 * f(x)| > s/2\}|$

$$\le 2^p s^{-p} \|K_1 * f\|_p^p \le 2^p s^{-p} \|K_1\|_1^p .$$

But, $\|K_1\|_1 = \int_{|x| \le q^k} |x|^{\text{Re}(\alpha)-n} dx = c_2\, q^{k\,\text{Re}(\alpha)}$.

Hence, $|\{x : |K * f(x)| > s\}| \le (2\, c_2\, s^{-1}\, q^{k\,\text{Re}(\alpha)})^p$

$$\le (2\, c_2)^p\, s^{-p}\, q^{\text{Re}(\alpha)p}\left(\frac{2\,c_1}{s}\right)^{r\,\text{Re}(\alpha)p/n}$$

$$= c_4\, s^{-p(1 + r\,\text{Re}(\alpha)/n)} = c_4\, s^{-r} .$$

This completes the proof.

We have obtained two differing interpretations of $I^{\alpha}\varphi$, $\varphi \in \mathscr{S}$ when $0 < \text{Re}(\alpha) < n$. For one, we have that $I^{\alpha}\varphi \in \mathscr{S}'$ defined by $(I^{\alpha}\varphi)\hat{} = |x|^{-\alpha}\hat{\varphi}$. For another, since $\varphi \in \mathscr{S} \subset L^p$, $I^{\alpha}\varphi$ is defined as an a.e. convergent integral that is in L^r for $r > (1 - \text{Re}(\alpha)/n)^{-1}$ and in this sense $I^{\alpha}\varphi \in \mathscr{S}'$ also. For $\varphi \in \mathscr{S}$ the two notions agree of course. For more general functions we have:

<u>Corollary (4.10)</u>. <u>If</u> $f \in L^p$, $(1/p) - (\text{Re}(\alpha)/n) = 1/r > 0$. $\text{Re}(\alpha) > 0$, $1 < p < \infty$, <u>then</u> $(I^{\alpha}f)\hat{} = |x|^{-\alpha}\hat{f}$ <u>in</u> $\mathscr{S}'$ <u>in the sense that if</u> $\{\varphi_k\}$ <u>is a sequence in</u> $\mathscr{S}$ <u>that converges to</u> f <u>in</u> L^p <u>then</u> $\{(I^{\alpha}\varphi_k)\hat{}\} = \{|x|^{-\alpha}\hat{\varphi}_k\} \to (I^{\alpha}f)\hat{}$ <u>in</u> $\mathscr{S}'$.

<u>Proof</u>. $\{I^{\alpha}\varphi_k\} \to I^{\alpha}f$ in L^r, from (4.9). Thus, $\{(I^{\alpha}\varphi_k)\hat{}\} \to (I^{\alpha}f)\hat{}$ in $\mathscr{S}'$. From (4.5) we see that $(I^{\alpha}\varphi_k)\hat{} = |x|^{-\alpha}\hat{\varphi}_k$.

We anticipate a result from Chapter IV on regularizations.

Lemma (4.11). *If* $1 \leq p < \infty$, $f \in L^p$ *or* $p = \infty$, f *uniformly continuous, then* $f(\cdot,k) = f * R(\cdot,k)$ *converges in* L^p *to* f, *as* $k \to -\infty$. *If* $\varphi \in \mathcal{S}$ *then* $\varphi(\cdot,k) \to \varphi$ *in* $\mathcal{S}$ *as* $k \to -\infty$. *If* $f \in \mathcal{S}'$ *then* $f(\cdot,k) \to f$ *in* $\mathcal{S}'$ *as* $k \to -\infty$. *Note also that* $f(\cdot,k) \in \mathcal{O}_M$.

Proof. Observe that $(f(\cdot,k))\hat{} = \hat{f}\Phi_k$. For $\varphi \in \mathcal{S}$, $\hat{\varphi}$ is supported on some $\mathfrak{P}^\ell$ so $\varphi(\cdot,k) = \varphi$ if $k \leq \ell$. For $f \in L^p$, or f uniformly continuous see the argument II.(6.8). For $f \in \mathcal{S}'$ note that $(f(\cdot,k))\hat{} = \hat{f}\Phi_k \in \mathcal{O}_C$ so $f \in \mathcal{O}_M$. Now fix $\varphi \in \mathcal{S}$ and we may suppose $\hat{\varphi}$ is supported on $\mathfrak{P}^\ell$. Then if $k \leq \ell$, $(f(\cdot,k)\hat{},\varphi) = (\hat{f}\Phi_k,\varphi) = (\hat{f},\Phi_k\varphi) = (\hat{f},\varphi)$, so $(f(\cdot,k))\hat{} \to \hat{f}$ in $\mathcal{S}'$, hence, $f(\cdot,k) \to f$ in $\mathcal{S}'$.

We now come to the big theorem.

Theorem (4.12). *If* $0 < \mathrm{Re}(\alpha)$, $\mathrm{Re}(\beta)$, $\mathrm{Re}(\alpha+\beta) < n$ *and* $\varphi \in \mathcal{S}$ *then* $I^\alpha(I^\beta\varphi) = I^\beta(I^\alpha\varphi) = I^{\alpha+\beta}\varphi$ *as a.e. convergent integrals and as elements of* $\mathcal{S}'$

Proof. Let $1/r = (1/p) - (\mathrm{Re}(\alpha)/n)$, $1/s = (1/r) - (\mathrm{Re}(\beta)/n)$ $= (1/p) - (\mathrm{Re}(\alpha+\beta)/n)$, where $p > 1$ is chosen so $s < \infty$. Let $\varphi_k(x) = \varphi(x,k)$, $\varphi \in \mathcal{S}$. Then $I^\alpha\varphi_k(x) = I^\alpha\varphi(x,k)$. (Note that $(I^\alpha\varphi)\hat{}_k(\xi)$ $= (|\xi|^{-\alpha}\hat{\varphi})\Phi_k = |\xi|^{-\alpha}(\hat{\varphi}\Phi_k) = |\xi|^{-\alpha}\hat{\varphi}_k = (I^\alpha(\varphi_k))\hat{}(\xi)$. $(I^\alpha\varphi_k)\hat{} \in \mathcal{O}_C$ so $I^\alpha\varphi_k \in \mathcal{O}_M$ and so $(I^\alpha\varphi_k)(x) = (I^\alpha\varphi)(x,k)$ as locally constant functions.)

This equality implies that the (a.e.) convergent integrals that represent the two functions are equal. a.e. Repeating the argument (with slight modification since $I^{\alpha}\varphi(x,k) \in \mathcal{O}_M$, but $\notin \mathcal{S}$) so we obtain $(I^{\alpha+\beta}\varphi)(x,k) = I^{\beta}((I^{\alpha}\varphi)_k)(x) = I^{\beta}(I^{\alpha}(\varphi_k))(x)$ as functions that are defined as a.e. convergent integrals that are in $\mathcal{O}_M \subset \mathcal{S}'$ and in L^s and hence they agree for all x. Furthermore Fourier transform of this function is $|x|^{-(\alpha+\beta)}\hat{\varphi}\,\Phi_k$, which is defined as a commutative associative product by (3.19).

On the right hand side we have $\varphi_k \to \varphi(L^p)$, so $I^{\alpha}(\varphi_k) \to I^{\alpha}\varphi(L^r)$ and $I^{\beta}(I^{\alpha}(\varphi_k)) \to I^{\beta}(I^{\alpha}\varphi)(L^s)$ by (4.11) and a double application of (4.9).

But $I^{\alpha+\beta}\varphi \in L^s$ by (4.9) and by (4.11) $I^{\alpha+\beta}\varphi(x,k) \to I^{\alpha+\beta}\varphi(x)(L^s)$. Thus, $I^{\alpha+\beta}\varphi(x) = I^{\beta}(I^{\alpha}\varphi)(x)$ as elements of L^s, as a.e. convergent integrals, etc.

5. Bessel potentials

The solution in the euclidean case to the problem of smoothing the kernels, $1/\Gamma_n(\alpha)|x|^{\alpha-n}$, whose Fourier transforms are $|x|^{-\alpha}$ is to consider distributions $G^{\alpha} \in \mathcal{S}'$ where $\hat{G}^{\alpha}(x) = (1+|x|^2)^{-\alpha/2}$. Our job is to find a suitable replacement. Notice that $(1+|x|^2)^{-\alpha/2}$ $= ((1+|x|^2)^{\frac{1}{2}})^{-\alpha} = |(1,x)|^{-\alpha}$ where $(1,x) \in R^{n+1}$, and $|(1,x)|$ is th norm in $\mathbb{R}^{n+1}$. An obvious choice for local fields is to try $G^{\alpha} \in \mathcal{S}'$ where $\hat{G}^{\alpha}(x) = (\max[1,|x|])^{-\alpha}$, since $\max[1,|x|] = |(1,x)|$ where $(1,x) = (1,x_1,\ldots,x_n) \in K^{n+1}$.

Notice that $\hat{G}^{\alpha} \in \mathcal{O}_M$ so G^{α} has compact support (namely $\mathfrak{D}$) $G^{\alpha} * f$ is well defined as an element of $\mathcal{S}'$ for all $\alpha \in C$, $f \in \mathcal{S}'$. It is completely trivial that as elements of $\mathcal{S}'$, whenever, $\alpha,\beta \in C$, $f \in \mathcal{S}'$ we have $G^{\beta}*(G^{\alpha}*f) = G^{\alpha}*(G^{\beta}f) = G^{\alpha+\beta}* f$, and $G^{\alpha}* G^{\beta} = G^{\alpha+\beta}$.

Definition. If $f \in \mathcal{S}'$, $\alpha \in C$ we define the <u>Bessel potential of order α</u> of f by

$$(J^{\alpha}f)^{\wedge} = (\max[1,|x|])^{-\alpha}\hat{f} .$$

$$G^{\alpha} \in \mathcal{O}_C \text{ is defined by } \hat{G}^{\alpha}(x) = (\max[1,|x|])^{-\alpha} .$$

Proposition (5.1). *For $\alpha,\beta \in C$, $f \in \mathcal{S}'$, we have that $J^{\alpha}f = G^{\alpha}* f \in \mathcal{S}'$. $J^{\alpha}(J^{\beta}f) = J^{\alpha+\beta}f$. The maps $f \to J^{\alpha}f$, $\varphi \to J^{\alpha}\varphi$ are homeomorphisms from $\mathcal{S}'$ onto $\mathcal{S}'$ and $\mathcal{S}$ onto $\mathcal{S}$ respectively. $(J^{\alpha})^{-1} = J^{-\alpha}$. Furthermore J^{α} is continuous in α in the sense that whenever $\{\alpha_k\} \to \alpha$ in C then $J^{\alpha_k}f \to J^{\alpha}f$ in $\mathcal{S}'$, when $f \in \mathcal{S}'$ and $J^{\alpha_k}\varphi \to J^{\alpha}\varphi$, when $\varphi \in \mathcal{S}$.*

Proof. Left as an easy exercise.

Our next result shows that if $\mathrm{Re}(\alpha) > 0$ then $G^{\alpha} \in L^1$.

Lemma (5.2). Suppose $\mathrm{Re}(\alpha) > 0$. Define K_α as follows:

If $\alpha \neq n$, $K_\alpha(x) = (1/\Gamma_n(\alpha))(|x|^{\alpha-n} - q^{\alpha-n})\Phi_0(x)$,

$$K_n(x) = (1-q^{-n})\log_q(q/|x|)\Phi_0(x) .$$

Then $K_\alpha \in L^1$ and $\hat{K}_\alpha(\xi) = (\max[1,|\xi|])^{-\alpha}$.

Remarks. It is obvious that $K_\alpha \in L^1$. Since $\hat{K}_\alpha = \hat{G}^\alpha$, we see that G^α is the function K_α and $G^\alpha \in L^1$. Thus, if $f \in L^p$, $1 \leq p \leq \infty$, $J^\alpha f(x) = G^\alpha * f(x) = \int G^\alpha(x-z)f(z)dz$, the integral converges a.e., $J^\alpha f \in L^p$, and $\|J^\alpha f\|_p \leq A_\alpha \|f\|_p$, $A_\alpha > 0$, where $A_\alpha = \|G^\alpha\|_1$.

Proof of Lemma (5.2). Assume $\mathrm{Re}(\alpha) > 0$, $\alpha \neq n$. Note that $\hat{\Phi}_0 = \Phi_0$

Thus,

$$\hat{K}_\alpha(\xi) = \begin{cases} -q^{\alpha-n}/\Gamma_n(\alpha), & |\xi| \leq 1 \\ 0 & , |\xi| > 1 \end{cases} + \frac{1}{\Gamma_n(\alpha)} \int_{|x|\leq 1} |x|^{\alpha-n}\bar{\chi}(\xi\cdot x)dx .$$

If $|\xi| \leq 1$, the second term is

$$1/\Gamma_n(\alpha) \int_{|x|\leq 1} |x|^{\alpha-n}dx = (1/\Gamma_n(\alpha))(1-q^n)/(1-q^{-\alpha}).$$

Thus, for $|\xi| \leq 1$, $\hat{K}_\alpha(\xi) = (1/\Gamma_n(\alpha))\left[\frac{1-q^{-n}}{1-q^{-\alpha}} - q^{\alpha-n}\right]$

$$= (1/\Gamma_n(\alpha))(1-q^{\alpha-n})/(1-q^{-\alpha}) = 1 .$$

For $|\xi| > 1$, we get

$$\hat{K}_\alpha(\xi) = (1/\Gamma_n(\alpha)) \int_{|x|\leq 1} \bar{\chi}(\xi\cdot x)|x|^{\alpha-n}dx$$

$$= |\xi|^{-\alpha}(1/\Gamma_n(\alpha)) \int_{|y|\leq |\xi|} \bar{\chi}(\xi'\circ y)|y|^{\alpha-n}dy$$

$$= |\xi|^{-\alpha}(1/\Gamma_n(\alpha))\Gamma_n(\alpha) = |\xi|^{-\alpha}, \quad \text{(by (4.2))}.$$

This proves the lemma for $\mathrm{Re}(\alpha) > 0$, $\alpha \neq n$. For $\alpha = n$, it is proved by a similar calculation, which is left as an exercise.

We now give several elementary properties of the functions G^α.

(5.3) $G^\alpha \in L^p(K^n)$ <u>if</u> $\mathrm{Re}(\alpha) > n/p'$, $1 \leq p \leq \infty$, $(1/p + 1/p' = 1)$. <u>In particular</u>, $G^\alpha \in L^1$ if $\mathrm{Re}(\alpha) > 0$, $G^\alpha \in L^\infty$ <u>if</u> $\mathrm{Re}(\alpha) > n$.

(5.4) $G^\alpha(x) \geq 0$, $|x| \neq 0$, <u>if</u> $\alpha > 0$.

(5.5) $\int G^\alpha(x)dx = 1$, <u>if</u> $\mathrm{Re}(\alpha) > 0$.

(5.3), (5.4) and (5.5) are immediate consequences of (5.2) and the remarks following its statement.

(5.6) $\{G^\alpha\}$ <u>is an analytic family of distributions in the sense that for every</u> $\varphi \in \mathcal{S}$, (G^α,φ) <u>is an entire function of</u> α.

<u>Proof</u>. We may suppose that $\hat{\varphi}$ is supported on $\mathfrak{B}^{-\ell}$, with $\ell \geq 0$. Then write $(G^{\alpha},\varphi) = ((\max[1,|x|])^{-\alpha},\hat{\varphi}) = \int_{x\in\mathfrak{D}} \hat{\varphi} + \int_{1<|x|\leq q^{\ell}} \hat{\varphi}(x)|x|^{-\alpha}dx$.

We complete this section with the simplest facts about the Lipschitz continuity of G^{α} in the L^p-norm.

(5.7) <u>If</u> $\mathrm{Re}(\alpha)-n/p' = \beta > 0$ <u>then</u> $\|G^{\alpha}(\cdot + h)-G^{\alpha}(\cdot)\|_p \leq A_{\alpha p}|h|^{\beta}$, <u>with</u> $A_{\alpha p} > 0$ <u>independent of</u> $h \in K^n$.

<u>Proof</u>. Since $\mathrm{Re}(\alpha)-n/p' = \beta > 0$, $G^{\alpha} \in L^p$ (from (5.3)) so if $|h| \geq 1$, $\|G^{\alpha}(\cdot + h)-G^{\alpha}\|_p \leq 2\|G^{\alpha}\|_p = (2\|G^{\alpha}\|_p |h|^{-\beta})|h|^{\beta}$, and for $|h| \geq 1$ $A_{\alpha p} \geq 2\|G^{\alpha}\|_p$ will do.

We now suppose $|h| \leq 1$. Since $|x+h| = |x|$ if $|x| > |h|$ and we know that G^{α} is radial (depends only on $|x|$) we see that

$$\|G^{\alpha}(\cdot + h) - G^{\alpha}(\cdot)\|_p = \begin{cases} \left[\int_{|x|\leq|h|} |G^{\alpha}(x+h)-G^{\alpha}(x)|^p dx\right]^{1/p}, & 1 \leq p < \infty \\ \sup_{|x|\leq|h|} |G^{\alpha}(x+h)-G^{\alpha}(x)|, & p = \infty \end{cases}.$$

Suppose $p = \infty$. Then $\mathrm{Re}(\alpha) > n$ so $\alpha \neq n$. Thus, $\Gamma_n(\alpha)G^{\alpha}(x) = -q^{\alpha-n}\Phi_0(x) + |x|^{\alpha-n}\Phi_0(x)$. Since $\Phi_0(x+h) = \Phi_0(x)$ for $|x| \leq |h| < 1$ (both terms are 1) we see that

$$|\Gamma_n(\alpha)|\ \|G^{\alpha}(\cdot + h) - G^{\alpha}(\cdot)\|_{\infty} \leq \sup_{|x|\leq |h|<1} \left||x+h|^{\alpha-n} - |x|^{\alpha-n}\right| \leq 2|h|^{\beta}.$$

This completes the proof for $p = \infty$.

Now assume that $1 \leq p < \infty$, $\alpha \neq n$, $|h| \leq 1$.

$$|\Gamma_n(\alpha)| \, \|G^{\alpha}(\cdot + h) - G^{\alpha}(\cdot)\|_p = \left[\int_{|x| \leq |h|} \left| |x+h|^{\alpha-n} - |x|^{\alpha-n} \right|^p \right]^{1/p}$$

$$\leq 2 \left[\int_{|x| \leq |h|} |x|^{\beta p} dx \right]^{1/p} = A_{\alpha p} |h|^{\beta} .$$

A single case is left: $1 \leq p < \infty$, $\alpha = n$.

We will find a constant $A_n > 0$ such that $\|G^n(\cdot+h) - G^n(\cdot)\|_p^p \leq A_n^p |h|^n$.

$$\|G^n(\cdot+h) - G^n(\cdot)\|_p^p = \int_{|x| \leq |h|} |G^n(x+h) - G^n(x)|^p dx$$

$$= \int_{|x| < |h|} |G^n(x+h) - G^n(x)|^p dx + \int_{|x+h| < |h| = |x|} |G^n(x+h) - G^n(x)|^p dx$$

$$+ \int_{|x+h| = |h| = |x|} |G^n(x+h) - G^n(x)|^p dx .$$

The first two terms are equal, and since G^n is radial, the third term is zero. Thus,

$$\|G^n(\cdot+h) - G^n(\cdot)\|_p^p = 2 \int_{|x| < |h|} |G^n(x+h) - G^n(x)|^p dx .$$

Set $|h| = q^{-\ell}$, $\ell \geq 0$. For $|x| < |h|$, $G^n(x+h) = G^n(h)$ $= (\ell+1)(1-q^{-n})$. For $|x| = q^{-k}$ $(k > \ell)$, $G^n(x) = (k+1)(1-q^{-n})$.

Thus, $|G^n(x+h) - G^n(x)| = (k-\ell)(1-q^{-n})$. Hence,

$$\int_{|x|<|h|} |G^n(x+h) - G^n(x)|^p dx = (1-q^{-n})^{p+1} \sum_{k=\ell+1}^{\infty} (k-\ell)^p q^{-kn}$$

$$= (1-q^{-n})^{p+1} q^{-\ell n} \sum_{k=\ell+1}^{\infty} (k-\ell)^p \, q^{-(k-\ell)n}$$

$$= (1-q^{-n})^{p+1} \Big(\sum_{k=1}^{\infty} k^p \, q^{-kn} \Big) |h|^n .$$ This completes the proof.

6. The relation of I^α and J^α

Notation. Suppose $f(x) = \begin{cases} g(x), & |x| \le 1 \\ h(x), & |x| > 1 \end{cases}$. Then we write $f(x) = \{g(x), h(x)\}$. For example $\hat{G}^\alpha(x) = \{1, |x|^{-\alpha}\}$.

Theorem (6.1). *For* $\operatorname{Re}(\alpha) > 0$ *there are finite Borel measures* $M_{1,\alpha}$ *and* $M_{2,\alpha}$ *such that*

(i) $|x|^\alpha = \hat{M}_{1,\alpha}(x)\, \hat{G}^{-\alpha}(x)$

(ii) $\hat{G}^{-\alpha} = |x|^\alpha + \hat{M}_{2,\alpha}(x)$.

Proof. $\hat{M}_{1,\alpha}(x) = \{|x|^\alpha, 1\} = 1 - \Phi_0 + \{|x|^\alpha, 0\}$, and $\hat{M}_{2,\alpha}(x) = \Phi_0 - \{|x|^\alpha, 0\}$. Hence the theorem follows if $\{|x|^\alpha, 0\}$ is the Fourier transform of a finite measure. Actually it is the Fourier transform of an L^1 function

An easy calculation shows that $\{|x|^{\alpha},0\}^{\wedge}(u)$ $= \{(1-q^{-n})/(1-q^{\alpha-n}),\ \Gamma(\alpha+n)|u|^{-(\alpha+n)}\}$. But this function is in L^1 and $\{|x|^{\alpha},0\}$ is continuous. From (1.16) it follows that $\{|x|^{\alpha},0\}$ is the Fourier transform of an L^1 function.

Definition. For $\alpha \in C$, $1 \leq p \leq \infty$, let $L^p_{\alpha} = \{f : f = J^{\alpha}g,\ g \in L^p\}$. We norm L^p_{α} with $\|f\|_{p\alpha} = \|J^{-\alpha}f\|_p$. We also let $C_{0,\alpha} = \{f : f = J^{\alpha}g,\ g \in C_0\}$. $C_{0,\alpha}$ is supplied with the L^{∞} norm.

Some Facts. L^p_{α} is isomorphic, as a Banach space with L^p, $C_{0,\alpha}$ with C_0, for all α. Clearly $\mathcal{S}$ is dense in L^p_{α}, $1 \leq p < \infty$ and in $C_{0,\alpha}$ for all α, as follows from the fact that if $\varphi \in \mathcal{S}$ then $J^{\alpha}\varphi \in \mathcal{S}$.

Note that if $f \in \mathcal{S}'$ then $\Phi_k(x)f(x,k) \to f$ in $\mathcal{S}'$, but $\Phi_k(x)f(x,k) \in \mathcal{S}$ for all k. Thus, $\mathcal{S}$ is dense in $\mathcal{S}'$. Thus, L^p_{α} and $C_{0\alpha}$ are dense in $\mathcal{S}'$ for every p and α. With respect to inclusions we see that $\mathcal{S} \subset L^p_{\alpha} \subset \mathcal{S}'$ for all p,α, $\mathcal{S} \subset C_{0,\alpha} \subset \mathcal{S}'$ for all α; the inclusions are continuous. Since $\hat{G}^{\alpha}$ is a radial function we see that G^{α} is radial and so $\tilde{G}^{\alpha} = G^{\alpha}$. As a consequence, J^{α} is self adjoint in the sense $(J^{\alpha}f,\varphi) = (f,J^{\alpha}\varphi)$ for all $f \in \mathcal{S}'$, $\varphi \in \mathcal{S}$, $\alpha \in C$. $(J^{\alpha}f,\varphi) = (G^{\alpha} * f,\varphi) = (f,\tilde{G}^{\alpha} * \varphi) = (f,G^{\alpha} * \varphi) = (f,J^{\alpha}\varphi)$.

<u>Corollary (6.2)</u>. <u>Suppose</u> $Re(\alpha) > 0$, $f \in \mathscr{S}$. <u>Let</u> $g = I^{-\alpha} f \in C_0 \cap L^2$ $(\hat{g} = |x|^{\alpha} \hat{f} \in L^1 \cap L^2)$. <u>Then</u> $g \in L^p$ <u>and there are positive constants</u> A_α, B_α <u>independent of</u> f <u>and</u> p <u>such that</u>

$$A_\alpha \|f\|_{p\alpha} \leq \|f\|_p + \|g\|_p \leq B_\alpha \|f\|_{p\alpha} .$$

<u>Proof</u>. There is an $h \in L^p$ such that $f = J^\alpha h = G^\alpha * h$. $(h = J^{-\alpha} f \in \mathscr{S} \subset L^p.)$ Hence, $\|f\|_{p\alpha} = \|h\|_p$; $\|f\|_p \leq \|G^\alpha\|_1 \, \|h\|_p = \|G^\alpha\|_1 \|f\|_{p\alpha}$. $\hat{g} = |x|^\alpha \hat{f} = |x|^\alpha / \{1, |x|^\alpha\} \, \hat{h} = \hat{M}_{1,\alpha} \, \hat{h}$. Thus, $g = M_{1,\alpha} * h$, $\|g\|_p \leq \|M_{1,\alpha}\|_M \, \|h\|_p$ with $B_\alpha = \|G^\alpha\|_1 + \|M_{1,\alpha}\|_M$, $\|f\|_p + \|g\|_p \leq B_\alpha \|f\|_{p\alpha}$.

On the other hand, $\hat{h} = \{1, |x|^\alpha\} \hat{f} = (|x|^\alpha + \hat{M}_{2,\alpha}) \hat{f} = \hat{g} + \hat{M}_{2,\alpha} \hat{f}$. Thus, $h = g + (M_{2,\alpha} * f)$, and so $\|f\|_{p\alpha} = \|h\|_p \leq \|g\|_p + \|M_{2,\alpha}\|_M \|f\|_p$ $\leq (1 + \|M_{2,\alpha}\|_M)(\|g\|_p + \|f\|_p)$. The result follows with $A_\alpha = 1/(1 + \|M_{2,\alpha}\|_M)$.

Notice that if $f \in L^p_\alpha$, $1 \leq p < \infty$ or $f \in C_{0\alpha}$, $p = \infty$, and $Re(\alpha) > 0$, then we can find a sequence $\{f_k\}$ in $\mathscr{S}$ such that $\{f_k\} \to f$ in L^p_α , $(C_{0,\alpha}$ respectively) and $\{I^{-\alpha} f_k\}$ is Cauchy in L^p, and we can define its limit in L^p to be $I^{-\alpha} f$. We can define $I^\alpha f$ in the following circumstances. If $f \in \mathscr{S}'$ and there is a sequence $\{f_k\}$ in $\mathscr{S}$, such that $\{I^\alpha f_k\}$ is Cauchy in L^p; its limit in L^p is defined to be $I^\alpha f$. Our considerations in this section show that a natural domain for $I^{-\alpha}$, $Re(\alpha) > 0$ is L^p_α , $1 \leq p < \infty$ and $C_{0,\alpha}$.

Corollary (6.3). *If* $\mathrm{Re}(\alpha) > 0$ *then* $f \in L^p_\alpha$ *iff* $f \in L^p$ *and* $I^{-\alpha} f \in L^p$. *Furthermore* $\|f\|_{p\alpha}$ *is equivalent to* $\|f\|_p + \|I^{-\alpha} f\|_p$.

7. Lebesgue spaces of Bessel potentials

We know that $G^\alpha \in L^r$ if $\mathrm{Re}(\alpha) > n/r'$. Suppose $f \in L^p_\alpha$. Then $f = g * G^\alpha$, $g \in L^p$. By Young's convolution theorem we know that if $0 \le 1/s = (1/p) + (1/r) - 1 \le 1$ then $f \in L^s$ and $\|f\|_s \le \|g\|_p \|G^\alpha\|_r = \|G^\alpha\|_r \|f\|_{p\alpha}$. This relation between p,r and s can be written, $(1/p) - (1/s) = 1 - (1/r) = 1/r'$. We obtain:

Lemma (7.1). *If* $1 \le p \le s \le \infty$, $\mathrm{Re}(\alpha) > n((1/p) - (1/s)) \ge 0$, *then* $L^p_\alpha(K^n) \subset L^s(K^n)$. *The inclusion is continuous.*

Suppose now that $\mathrm{Re}(\alpha) = n((1/p) - (1/s)) > 0$, $1 < p < s < \infty$. Then it is easy to see that for such α, $|G^\alpha(x)| \le A_\alpha |x|^{\mathrm{Re}(\alpha) - n}$, $A_\alpha > 0$ and independent of x. As a consequence of (4.9) we have, with $B_\alpha = A_\alpha |1/\Gamma_n(\alpha)|$, $\|f\|_s = \|J^\alpha g\|_s \le B_\alpha \|I^{\mathrm{Re}(\alpha)} g\|_s \le B_\alpha C_{ps} \|g\|_p = B_\alpha C_{ps} \|f\|_{p\alpha}$. Thus,

Lemma (7.2). *If* $1 < p < s < \infty$, $\mathrm{Re}(\alpha) = n((1/p) - (1/s)) > 0$ *then* $L^p_\alpha(K^n) \subset L^s(K^n)$. *The inclusion is continuous.*

In the remainder of this section we will be concerned with showing the conclusion of (7.1) and (7.2) in the case $1 < p = s < \infty$, $\mathrm{Re}(\alpha) = 0$. In a later section, V §1, we will show that this fact is a trivial consequence of a multiplier theorem and depends entirely on the fact that $\hat{G}^{\alpha}(x)$ is bounded and radial, when $\mathrm{Re}(\alpha) = 0$. The techniques used here foreshadow methods that will be systematically developed later and are presented here for historical and pedagogical reasons.

Lemma (7.3). (A case of the Marcinkiewicz theorem)
Suppose T *is a linear operator defined on* $\mathcal{S}$ *with values in* $\mathcal{S}'$. *Suppose further that when* $f \in \mathcal{S}$ *then* $Tf \in L^2$ *and* $\|Tf\|_2 \leq A_2\|f\|_2$, $A_2 > 0$ *independent of* f; *and* Tf *is a function such that for all* $s > 0$, $|F_s| = |\{x : |Tf(x)| > \}| \leq A_1\|f\|_1 s^{-1}$, $A_1 > 0$ *independent of* f *and* s. *Then there are constants* $A_p > 0$ *independent of* f, $1 < p \leq 2$ *such that for* $f \in \mathcal{S}$, $Tf \in L^p$ *and* $\|Tf\|_p \leq A_p\|f\|_p$.

Proof. In proof of the Marcinkiewicz interpolation theorem we use decompositions of the form $f = f_{s1} + f_{s2}$ where $f_{s1}(x) = f(x)$ when $|f(x)| \leq \mu(s)$, and is zero otherwise, where $\mu(s)$ is chosen in the proof. Notice that if $f \in \mathcal{S}$ then $f_{s2} \in \mathcal{S}$. Reference to any standard proof of the interpolation theorem will now suffice.

<u>**Definition.**</u> An operator T that is defined on $\mathcal{S}$ with values in $\mathcal{S}'$ is said to <u>commute with translations</u> if for all $h \in K^n$, $\tau_h T = T\tau_h$.

<u>**Lemma (7.4).**</u> <u>If</u> T <u>is a linear operator defined on</u> $\mathcal{S}$ <u>with values in</u> $\mathcal{S}'$ <u>that commutes with translations then for all</u> $\varphi, \psi \in \mathcal{S}$, $(T\varphi, \psi) = (T\tilde{\psi}, \tilde{\varphi})$.

<u>Proof</u>. Since T is linear we may assume that $\varphi = \tau_u \Phi_k$, $\psi = \tau_v \Phi_k$ for some $k \in \mathbb{Z}$, $u,v \in K^n$. Note that $\tilde{\varphi} = \tau_{-u}\Phi_k$, $\tilde{\psi} = \tau_{-v}\Phi_k$. Then

$(T\varphi, \psi) = (T(\tau_u \Phi_k), \tau_v \Phi_k) = (\tau_u T\Phi_k, \tau_v \Phi_k) = (\tau_{-v} T\Phi_k, \tau_{-u}\Phi_k)$

$= (T(\tau_{-v}\Phi_k), \tau_{-u}\Phi_k) = (T\tilde{\psi}, \tilde{\varphi})$.

<u>**Lemma (7.5).**</u> (Corollary to (7.4)). <u>Suppose</u> T <u>is a linear operator defined on</u> $\mathcal{S}$ <u>with values in</u> $\mathcal{S}'$ <u>that commutes with translations</u>. <u>Suppose further that</u> $\varphi \in \mathcal{S}$ <u>implies that</u> $T\varphi \in L^r$, $\|T\varphi\|_r \le A\|\varphi\|_p$, $A > 0$ <u>independent of</u> φ , $1 \le p,r \le \infty$. <u>Let</u> $(1/p) + (1/p')$ $= (1/r) + (1/r') = 1$. <u>Then</u> $T\varphi \in L^{p'}$ <u>for all</u> $\varphi \in \mathcal{S}$ <u>and</u> $\|T\varphi\|_{p'} \le A\|\varphi\|_{r'}$.

<u>Proof</u>. If $\varphi, \psi \in \mathcal{S}$, $|(T\varphi, \psi)| = |(T\tilde{\psi}, \tilde{\varphi})| = \left|\int (T\tilde{\psi})(x)\, \tilde{\varphi}(x)dx\right|$ $\le \|T\tilde{\psi}\|_r \, \|\tilde{\varphi}\|_{r'} \le A\|\tilde{\psi}\|_p \, \|\tilde{\varphi}\|_{r'} = A\|\psi\|_p \, \|\varphi\|_{r'}$. Thus, the map $\psi \longrightarrow (T\varphi, \psi)$ is a bounded linear functional on $\mathcal{S} \cap L^p$ with norm

bounded by $A\|\varphi\|_{r'}$. Hence there is a $g \in L^{p'}$, $\|g\|_{p'} \leq A\|\varphi\|_{r'}$ such that $(T\varphi, \psi) = \int g(x)\psi(x)dx$ for all $\psi \in \mathcal{S}$. Thus $T\varphi \in L^{p'}$ and $\|T\varphi\|_{p'} = \|g\|_{p'} \leq A\|\varphi\|_{r'}$.

Lemma (7.6). *There are constants* $A, B > 0$, *depending only on* K^n, *such that if* $f \in L^1(K^n)$, $f(x) \geq 0$, $s > 0$ *then there is a countable collection of mutually disjoint spheres* $\{\omega_t\}_{t=1}^{\infty}$, $D_s = \cup_t \omega_t$, *and a decomposition of* f, $f = f_1^s + f_2^s$ *such that*:

(i) $|D_s| = \sum_t |\omega_t| < \|f\|_1 s^{-1}$

(ii) $|f(x)| \leq s$ a.e. *for* $x \notin D_s$

(iii) $|f_2^s(x)| \leq q^n s$ a.e., $x \in D_s$

(iv) $f_2^s(x) = f(x)$, $x \notin D_s$

(v) $f_1^s(x) = 0$, $x \notin D_s$

(vi) $\int_{\omega_t} f_1^s(x)dx = 0$, *for all* t.

Proof. If ω is a sphere, $\omega = y + \mathfrak{P}^k$ we set $\omega^* = y + \mathfrak{P}^{k-1}$. Thus, $\omega \subset \omega^*$, $|\omega| q^n = |\omega^*|$ and there are q^n distinct spheres ω' such that $(\omega')^* = \omega^*$.

Since $f \in L^1$ we can find $k \in \mathbb{Z}$ so large that if $|\omega| = q^{kn}$ then $\frac{1}{|\omega|}\int_{\omega} f(x)dx < s$. We will find a collection of mutually disjoint spheres $\{\omega_t\}$ such that $\frac{1}{|\omega_t|}\int_{\omega_t} f(x)dx \geq s$, but $\frac{1}{|\omega_t^*|}\int_{\omega_t^*} f(x)dx < s$. From this last inequality, the fact that $\omega_t \subset \omega_t^*$ and $q^n|\omega_t| = |\omega_t^*|$ we obtain

$$s \leq \frac{1}{|\omega_t|}\int_{\omega_t} f(x)dx = \frac{q^n}{|\omega_t^*|}\int_{\omega_t} f(x)dx \leq q^n \frac{1}{|\omega_t|^*}\int_{\omega_t^*} f(x)dx < q^n s \,.$$

Thus,

$$s \leq \frac{1}{|\omega_t|}\int_{\omega_t} f(x)dx < q^n s \quad . \tag{7.7}$$

The ω_t are chosen as follows: We start with the ω's above where $\frac{1}{|\omega|}\int_{\omega} f(x)dx < s$. Each ω is partitioned into the q^n spheres ω' such that $(\omega')^* = \omega$. If $\frac{1}{|\omega'|}\int_{\omega'} f(x)dx \geq s$, then ω' is one of our spheres. If $\frac{1}{|\omega'|}\int_{\omega'} f(x)dx < s$, then we continue by partitioning ω'. In this manner we get a countable collection $\{\omega_t\}$, that satisfies (7.7), and $D_s = \cup_t \omega_t$.

Now let $f_2^s(x) = \begin{cases} f(x), & x \notin D_s \\ \frac{1}{|\omega_t|}\int_{\omega_t} f(x)dx, & x \in \omega_t \end{cases}$, $f_1^s = f - f_2^s$.

(iv) and (v) are trivially verified. If $x \in \omega_t$,

$f_1^s(x) = f(x) - \frac{1}{|\omega_t|}\int_{\omega_t} f(z)dz = \frac{1}{|\omega_t|}\int_{\omega_t}(f(x) - f(z))dz$. We see that (vi) is valid.

If $x \notin D_s$ then for all ω, $x \in \omega$, $\frac{1}{|\omega|}\int f(z)dz < s$. It follows from (1.14) that for a.e. $x \notin D_s$, $f(x) \leq s$. This is (ii). For $x \in$ $f_s^s(x) = \frac{1}{|\omega_t|}\int_{\omega_t} f(x)dx < q^n s$ and this is (iii).

For each t, $s|\omega_t| \leq \int_{\omega_t} f(x)dx < q^n s|\omega_t|$. Thus,

$$s|D_s| \leq \int_{D_s} f(x) < q^n\, s|D_s| \quad . \tag{7.8}$$

In particular, $s|D_s| \leq \int f = \|f\|_1$ and (i) is established. This completes the proof.

<u>Remarks (7.9)</u>. <u>The conclusions of</u> (7.6) <u>yield the following results:</u>

(a) $f_1^s \in L^1$, $\|f_1^s\|_1 \leq 2\|f\|_1$.

(b) $f_2^s \in L^1$, $\|f_2^s\|_1 = \|f\|_1$; $f_2^s \in L^\infty$, $\|f_2^s\|_\infty \leq q^n s$; $f_2^s \in L^2$, $\|f_2^s\|_2 \leq (s\, q^n\|f\|_1)^{\frac{1}{2}}$.

<u>Proof.</u> $f_2^s \in L^\infty$, $\|f_2^s\|_\infty \leq q^n s$ follows from (ii), (iii) and (iv). $f_2^s \in L^1$, $\|f_2^s\|_1 = \|f\|_1$ is a trivial computation. Note that, $\|f_2^s\|_2 \leq \|f_1^s\|_1^{\frac{1}{2}} \cdot \|f_2^s\|_\infty^{\frac{1}{2}}$, and $\|f_1^s\|_1 \leq \|f_2^s\|_1 + \|f\|_1$. Done.

Lemma (7.10). For $\alpha \in C$, $\alpha \neq 0$, $n, k \in \mathbb{Z}$, $k \leq 0$,

$$G^{\alpha}(x,k) = \begin{cases} [(1-q^{-n})/(1-q^{\alpha-n})]\, q^{k(\alpha-n)} - (q^{\alpha-n}/\Gamma_n(\alpha)), & |x| \leq q^k \\ (|x|^{\alpha-n} - q^{\alpha-n})/\Gamma_n(\alpha), & q^k < |x| \leq 1 \\ 0, & |x| > 1 \end{cases}$$

$$G^{\alpha}(\cdot,k) \in L^p(K^n),\ 1 \leq p \leq \infty,\ k \leq 0 .$$

Proof. By observation the functions are in L^p, $1 \leq p \leq \infty$. They are also radial and continuous. A trivial computation shows that the functions have Fourier transforms $\{1,|x|^{-\alpha}\}\Phi_k(x)$ so that they are indeed, $G^{\alpha}(x,k)$.

Remarks (7.11). The cases $\alpha = 0$ and $\alpha = n$ are easily supplied.

(a) $G^{0}(x,k) = R(x,k)$

(b) $$G^{n}(x,k) = \begin{cases} 1-k(1-q^{-n}), & |x| \leq q^k \\ (1-q^{-n})\log_q(q/|x|), & q^k < |x| \leq 1 \\ 0, & |x| > 1 \end{cases} ,\quad k \leq 0 .$$

Lemma (7.12). Suppose $\mathrm{Re}(\alpha) = 0$.

(a) If $f \in \mathcal{S}'$, $G^{\alpha}(\cdot,k) * f \to J^{\alpha}f$ in $\mathcal{S}'$ as $k \to -\infty$.

(b) If $f \in L^2$, $G^{\alpha}(\cdot,k) * f \to J^{\alpha}f$ in L^2 as $k \to -\infty$, $\|J^{\alpha}f\|_2 = \|f\|_2$.

(c) If $f \in L^1$, $G^{\alpha}(\cdot,k) * f$ converges a.e. (as $k \to -\infty$) to a function f_{α} (not necessarily a distribution of function type). There is a constant $A > 0$ independent of f and $s > 0$ such that

$|\{x : |f_\alpha(x)| > s\}| \le A\|f\|_1 s^{-1}$.

<u>Proof</u>. If $f \in \mathscr{S}'$ then $(G^\alpha(\cdot,k) * f)\hat{} = (\hat{G}^\alpha \Phi_k)\hat{f} = (\hat{G}^\alpha \hat{f})\Phi_k = (J^\alpha f(\cdot,k$
Thus $G^\alpha(\cdot,k) * f = J^\alpha f(\cdot,k) \to J^\alpha f$ in $\mathscr{S}'$. (We used (3.19) and (4.11)

If $f \in L^2$, then $G^\alpha(\cdot,k) * f \in L^2$, since $G^\alpha(\cdot,k) \in L^1$. $(G^\alpha(\cdot,k) * f)\hat{} = \Phi_k(x)\{1,|x|^{-\alpha}\}\hat{f}(x)$. Note that not only is $J^\alpha f \in L^2$, $\|J^\alpha f\|_2 = \|\hat{G}^\alpha \hat{f}\|_2 = \|\hat{f}\|_2 = \|f\|_2$, but $f \to J^\alpha f$ is unitary when $\operatorname{Re}(\alpha) = 0$. Thus, $\|G^\alpha(\cdot,k) * f - J^\alpha f\|_2 = \left[\int_{|x| > q^{-k}} |\hat{f}(x)|^2 dx\right]^{\frac{1}{2}} = o$ as $k \to -\infty$. Furthermore $G^\alpha(\cdot,k) * f(x) \to J^\alpha f(x)$ a.e.

Now suppose $f \in L^1$. We may assume that $f(x) \ge 0$, $f \in L^1$ and $s > 0$ is given. If $\alpha = 0$ the result is immediate, since $G^0(\cdot,k) * f = R(\cdot,k) * f = f(x,k) \to f(x)$ a.e. (by (4.11)) and the fac that $|\{x : f(x) > s\}| \le \|f\|_1 s^{-1}$. That is $f_0 = f$.

We now suppose $\operatorname{Re}(\alpha) = 0$, $\alpha \ne 0$. Write $f = f_1^s + f_2^s$ as in (7.6) - (7.9). Note that $f_2^s \in L^2$, $\|f_2^s\|_2^2 \le q^n s\|f\|_1$. Thus, $(f_2^s)_\alpha($ exists a.e., $\|(f_2^s)_\alpha\|_2 = \|J^\alpha f_2^s\|_2 = \|f_2^s\|_2$. Hence,

$|\{x : |(f_2^s)_\alpha(x)| > s/2\}| \le \|(f_2^s)_\alpha\|_2^2\, 4 s^{-2} \le 4q^n \|f\|_1 s^{-1}$.

<u>Claim</u>. If $x \notin D_s$, $(f_1^s)_\alpha(x) = 0$.

<u>Proof of Claim</u>. $(f_1^s * G^\alpha(\cdot,k))(x) = \int f_1^s(x-z)G^\alpha(z,k)\,dz$, which is defined and continuous for all x since $f_1^s \in L^1$, $G^\alpha(\cdot,k) \in L^\infty$,

$G^{\alpha}(z,k) = 0$ if $z \notin \mathfrak{O}$. Thus, $(f_1^s * G^{\alpha}(\cdot,k))(x)$

$$= \int_{z\in\mathfrak{O}} f_1^s(x-z)G^{\alpha}(z,k)\,dz = \sum_t \int_{z\in(x-\omega_t)\cap\mathfrak{O}} f_1^s(x-z)G^{\alpha}(z,k)\,dz \quad .$$

Let us note that $0 \notin x - \omega_t$. If $0 \in x - \omega_t$ then $x - \omega_t = \mathfrak{P}^k$ for some k and $x \in \omega_t$, a contradiction since $x \notin D_s$. There are then two possibilities: $(x-\omega_t) \cap \mathfrak{O} = \emptyset$ and the term is zero, or $x-\omega_t \subset \mathfrak{O}$, $0 \notin x-\omega_t$, and $(x-\omega_t) \cap \mathfrak{O} = x-\omega_t$. In the latter case, each element in $x-\omega_t$ has the same norm, so $G^{\alpha}(z,k)$ is constant on the set. From the observation that $\int_{z\in(x-\omega_t)\cap\mathfrak{O}} f_1^s(x-z)\,dz = \int_{x-\omega_t} f_1^s(x-z)\,dz$ $= \int_{\omega_t} f_1^s(z)\,dz = 0$, we see that the remaining terms are also zero, so $(G^{\alpha}(\cdot,k) * f_1^s)(x) = 0$ for all $x \notin D_s$, $k \leq 0$, so $(f_1^s)_{\alpha}(x) = 0$, $x \notin D_s$. This proves the claim.

In particular $(f_1^s)_{\alpha}(x)$ exists if $x \notin D_s$ and so $f_{\alpha}(x)$ exists for a.e. $x \notin D_s$. But $|D_s| = o(1)$ as $s \to \infty$, so $f_{\alpha}(x)$ exists for a.e. x. Now we have,

$$|\{x : |f_{\alpha}(x)| > s\}| \leq |\{x : |(f_1^s)_{\alpha}(x)| > s/2\}| + |\{x : |(f_2^s)_{\alpha}(x)| > s/2\}|$$

$$\leq |D_s| + 4q^n\|f\|_1\, s^{-1} < \|f\|_1\, s^{-1} + 4q^n\|f\|_1\, s^{-1}$$

$$= (1 + 4q^n)\|f\|_1\, s^{-1} \; .$$

<u>Lemma (7.13)</u>. <u>If</u> $f \in L^p$, $1 < p < \infty$, $\text{Re}(\alpha) = 0$, <u>there is a constant</u> $A_p > 0$ <u>independent of</u> f <u>and</u> α <u>such that</u> $J^\alpha f \in L^p$ <u>and</u>
$\|J^\alpha f\|_p \leq A_p \|f\|_p$.

<u>Proof</u>. If $\alpha = 0$, $J^\alpha f = f$ and the result holds trivially. If $\alpha \neq$ we use (7.12). Consider the linear operator, T, defined by $T\varphi = \lim_{k \to -\infty} G^\alpha(\cdot,k) * \varphi = \varphi_\alpha = J^\alpha \varphi$. T maps $\mathscr{S}$ onto $\mathscr{S}$ and so into $\mathscr{S}'$. T commutes with translations:

$$(\tau_h(J^\alpha\varphi))^\wedge = \bar{\chi}_h(J^\alpha\varphi)^\wedge = \bar{\chi}_h(\hat{G}^\alpha\hat{\varphi}) = \hat{G}^\alpha(\bar{\chi}_h\hat{\varphi}) = \hat{G}^\alpha(\tau_h\varphi)^\wedge = (J^\alpha(\tau_h\varphi))^\wedge .$$

From (7.12)(b) $T\varphi \in L^2$, $\|T\varphi\|_2 = \|\varphi\|_2$. From (7.12) (c) there is a constant $A > 0$, independent of φ, α, $s > 0$, such that $|\{x: |T\varphi(x)| > s\}| \leq A\|\varphi\|_1 s^{-1}$. Applying (7.3) we obtain that there are constants $A_p > 0$, $1 < p \leq 2$, such that if $\varphi \in \mathscr{S}$, $T\varphi \in L^p$, an $\|T\varphi\|_p \leq A_p\|\varphi\|_p$. We now use (7.5) and obtain that there are constan $A_p > 0$, $1 < p < \infty$ $(A_p = A_{p'})$ such that whenever $\varphi \in \mathscr{S}$, $T\varphi \in L^p$, $\|T\varphi\|_p \leq A_p\|\varphi\|_p$.

Since $\mathscr{S}$ is dense in L^p we may choose a sequence $\{f_k\}$ in $\mathscr{S}$ that converges in L^p to $f \in L^p$. Since $\{f_k\} \to f$ in $\mathscr{S}'$, $\{J^\alpha f_k\} \to J^\alpha f$ in $\mathscr{S}'$. From the first part of the proof $\{J^\alpha f_k\}$ is Cauchy in L^p. Thus, there is a $g \in L^p$ such that $\{J^\alpha f_k\} \to g$ in

L^p and a portion in $\mathscr{S}'$. Thus $J^{\alpha}f = g \in L^p$. Furthermore,

$$\|J^{\alpha}f\|_p = \|g\|_p = \lim_k \|J^{\alpha}f_k\|_p \leq A_p \lim_k \|f_k\|_p = A_p\|f\|_p .$$

__Theorem (7.14).__ (a) __If__ $1 < p \leq r < \infty$, $\mathrm{Re}(\alpha) - n/p \geq \mathrm{Re}(\beta) - n/r$, __then__ $L^p_{\alpha}(K^n) \subset L^r_{\beta}(K^n)$.

(b) __If__ $1 \leq p \leq r \leq \infty$, $\mathrm{Re}(\alpha) - n/p > \mathrm{Re}(\beta) - n/r$ __then__ $L^p_{\alpha}(K^n) \subset L^r_{\beta}(K^n)$.

__The inclusion maps are continuous.__

__Proof.__ Since J^{β} is a linear isometry of L^r_{α} onto $L^r_{\alpha+\beta}$ we may assume that $\beta = 0$. Then (a) is a consequence of (7.2) and (7.13). (b) is a consequence of (7.1).

8. An elementary aspect of the Lipschitz theory

__Definition.__ If $\alpha > 0$, $\Lambda^p_{\alpha}(K^n) \equiv \Lambda^p_{\alpha}$ is that subset of L^p such that,

$$\sup_{h \in K^n} |h|^{-\alpha}\|f(\cdot + h) - f(\cdot)\|_p = \| \, |h|^{-\alpha}(f(x+h) - f(x))\|_{p\,\infty} < \infty .$$

We provide Λ^p_{α} with a norm $\|\cdot\|_{\alpha;p}$, where

$$\|f\|_{\alpha;p} = \|f\|_p + \| \, |h|^{-\alpha}(f(x+h) - f(x))\|_{p\,\infty} .$$

It is easy to verify that Λ^p_{α} is a Banach space.

<u>Theorem (8.1).</u> <u>If</u> $\mathrm{Re}(\alpha) > 0$, $1 \le p \le \infty$, <u>then</u> $L^p_\alpha \subset \Lambda^p_{\mathrm{Re}(\alpha)}$. <u>The inclusion is continuous.</u>

<u>Proof.</u> If $f \in L^p_\alpha$, $\mathrm{Re}(\alpha) > 0$, then $f \in L^p$ and $\|f\|_p \le A_\alpha \|f\|_{p\alpha}$, since $f = G^\alpha * g$, $g \in L^p$, $G^\alpha \in L^1$, with $\|g\|_p = \|f\|_{p\alpha}$, (see (5.3)). From (5.7) we have $\|G^\alpha(\cdot + h) - G^\alpha(\cdot)\|_1 \le A'_\alpha |h|^{\mathrm{Re}(\alpha)}$.

$f(x+h) - f(x) = g * (G^\alpha(\cdot + h) - G^\alpha(\cdot))(x)$, and so

$\|f(\cdot + h) - f(\cdot)\|_p \le \|g\|_p \|G^\alpha(\cdot + h) - G^\alpha(\cdot)\|_1 \le A'_\alpha \|f\|_{p\alpha} |h|^{\mathrm{Re}(\alpha)}$.

Thus, $\|f\|_{\alpha;p} \le (A_\alpha + A'_\alpha)\|f\|_{p\alpha}$.

9. Duals of the L^p_α spaces

<u>Notation.</u> If B is a Banach space we denote its dual by B^*.

<u>Notation.</u> Let M be the space of finite Borel measures with the norm $\|\mu\|_M \equiv$ total variation of μ. M is a Banach space. For $\alpha \in C$, we also define $M_\alpha = \{f = J^\alpha \mu : \mu \in M\}$. M_α is supplied with the norm $\|f\|_{M_\alpha} = \|J^{-\alpha} f\|_M$.

Since L^p_α, $1 \le p < \infty$; and $C_{0\alpha}$ are all Banach spaces that are linearly isometric with L^p, $1 \le p < \infty$; and C_0 respectively, it is obvious that the duals, as Banach spaces, are $L^{p'}$ and M respectivel

The point of the next result is to make this identification in a more natural way.

<u>Theorem (9.1)</u>. <u>Let</u> $\alpha \in C$. <u>Let</u> T <u>be a bounded linear functional on</u> L^p_α, $1 \le p < \infty$ <u>or on</u> $C_{0,\alpha}$ <u>for</u> $p = \infty$. <u>Then for</u> $\varphi \in \mathcal{S}$, $\varphi \to T\varphi \in \mathcal{S}'$. <u>That is</u>, $T\varphi = (u_T,\varphi)$ <u>for some</u> $u_T \in \mathcal{S}'$. <u>Furthermore</u>, $u_T \in L^{p'}_{-\alpha}$, $1 \le p < \infty$, $u_T \in M_{-\alpha}$ <u>for</u> $p = \infty$; <u>and the norm of</u> u_T <u>in</u> $L^{p'}_{-\alpha}$, M_α <u>respectively, is the norm of</u> T <u>as a linear functional</u>.

<u>The correspondence</u> $T \longleftrightarrow u_T$ <u>is a linear isometry of</u> $(L^p_\alpha)^*$ (<u>respectively</u> $(C_{0\,\alpha})^*$) <u>onto</u> $L^{p'}_{-\alpha}$ ($M_{-\alpha}$ <u>respectively</u>).

<u>Proof</u>. For notational convenience we will restrict attention to L^p_α, $1 \le p < \infty$. Suppose $f \to Tf$ is a bounded linear functional on L^p_α. Let its norm be A. Let $\{\varphi_k\}$ be a null sequence in $\mathcal{S}$. Then $\{J^{-\alpha}\varphi_k\}$ is also a null sequence in $\mathcal{S}$, so $\{J^{-\alpha}\varphi_k\}$ is a null sequence in L^p and so $\{\varphi_k\} \to 0$ in L^p_α. Hence $\varphi \to T\varphi \in \mathcal{S}'$. Let $u_T \in \mathcal{S}'$ be chosen so $T\varphi = (u_T, \varphi)$, $\varphi \in \mathcal{S}$.

Since $\mathcal{S}$ is dense in L^p, $\varphi \to (J^\alpha u_T, \varphi)$ can be extended to a linear functional on L^p with norm A, we see that there is a function $g \in L^{p'}$ such that $\|g\|_{p'} = A$ and $(J^\alpha u_T,\varphi) = \int g(x)\varphi(x)$ for all $\varphi \in \mathcal{S}$.

[Note that if $\varphi \in \mathscr{S}$, $|(J^{\alpha}u_T,\varphi)| = |(u_T, J^{\alpha}\varphi)|$
$\leq A\|^{\alpha}\varphi\|_{p,\alpha} = A\|\varphi\|_p$.]

Thus, $J^{\alpha}u_T = g \in L^{p'}$ and $u_T = J^{-\alpha}g \in L^{p'}_{-\alpha}$, $\|u_T\|_{p',-\alpha} = \|g\|_{p'} = A$.

On the other hand if we are given $u \in L^{p'}_{-\alpha}$, the map $f \to u(f) = \int (J^{\alpha}u)(x)(J^{-\alpha}f)(x)dx$ is a bounded linear functional on L^p_{α}, since $|u(f)| \leq \|J^{\alpha}u\|_{p'}\|J^{-\alpha}f\|_p = \|u\|_{p',-\alpha}\|f\|_{p,\alpha}$. The norm of $f \to u(f)$ is bounded by $\|u\|_{p',-\alpha}$. By the converse to Hölder's inequality it is equal to $\|u\|_{p',-\alpha}$.

10. Operators that commute with translation

Proposition (10.1). *For* $f \in \mathscr{S}'$ *let* $T\varphi = f * \varphi$, $\varphi \in \mathscr{S}$. *Then for all* $h \in K^n$, $\tau_h T = T\tau_h$.

Proof. Note that $\bar{x}_h \in \Theta_M$. Thus for $\varphi \in \mathscr{S}$, $(\tau_h(T\varphi))\hat{}$
$= \bar{x}_h(T\varphi)\hat{} = \bar{x}_h(\hat{f}\hat{\varphi}) = \hat{f}(\bar{x}_h\hat{\varphi}) = \hat{f}(\tau_h\varphi)\hat{} = (T(\tau_h\varphi))\hat{}$.

Lemma (10.2). $f \in L^p$, $g \in L^{p'}$, $(1/p) + (1/p') = 1$ *then* $f * g \in C_0$ *and* $\|f * g\|_{\infty} \leq \|f\|_p \|g\|_{p'}$.

Proof. The proof is standard. It depends on the fact that if $1 \leq p < \infty$, and $f \in L^p$, $\|f(\cdot + h) - f(\cdot)\|_p = o(1)$ as $|h| \to 0$.

Lemma (10.3). (Soboleff) *If* $f \in L^p_\alpha(K^n)$, $\mathrm{Re}(\alpha) > n/p$, *then* f *is a continuous function that is bounded by* $A\|f\|_{p\alpha}$ *where* $A > 0$ *is independent of* f.

Proof. From (7.14)(b) $L^p_\alpha \subset L^\infty_{\alpha-n/p}$ and $f \in L^p_\alpha$ implies $\|f\|_{\infty,\alpha-n/p} \le A_{p\alpha}\|f\|_{p\alpha}$. But $f \in L^\infty_{\alpha-n/p}$ implies $f = G^{\alpha-n/p} * g$, $g \in L^\infty$, $G^{\alpha-n/p} \in L^1$. Thus, $f \in C_o \subset L^\infty$, and $\|f\|_\infty \le \|G^{\alpha-n/p}\|_1\|g\|_\infty$ $= \|G^{\alpha-n/p}\|_1 \|f\|_{\infty,\alpha-n/p} \le \|G^{\alpha-n/p'}\|_1 A_{p\alpha}\|f\|_{p\alpha}$.

Theorem (10.4). *If* T *is a bounded linear operator from* $L^p(K^n)$ *to* $L^r(K^n)$ *for* $1 \le p, r \le \infty$ *and* T *commutes with translations, then there is a unique* $u \in \mathscr{S}'$ *such that for all* $\varphi \in \mathscr{S}$, $T\varphi = u * \varphi$.

Proof. (a) For $\alpha \in C$, $J^\alpha T = TJ^\alpha$ as operators on $\mathscr{S}$ into $\mathscr{S}'$: Note that if $f \in \mathscr{S}'$, $\varphi \in \mathscr{S}$ then $(J^\alpha f,\varphi) = (f,J^\alpha\varphi)$ and that $(J^\alpha\varphi)^\sim = J^\alpha(\tilde\varphi)$. Also note that TJ^α commutes with translations and is linear. Thus for $f \in \mathscr{S}'$, $\varphi, \psi \in \mathscr{S}$,

$$(J^\alpha T\varphi,\psi) = (T\varphi,J^\alpha\psi) = (T(J^\alpha\psi)^\sim,\tilde\varphi)$$

$$= (T(J^\alpha(\tilde\psi)),\tilde\varphi) = (TJ^\alpha(\tilde\psi),\tilde\varphi) = (TJ^\alpha\varphi,\psi)$$

(b) $\alpha \in C$, $\varphi \in \mathscr{S}$ implies $T\varphi \in L^r_\alpha$: $J^{-\alpha}\varphi \in \mathscr{S} \subset L^p$, so $J^{-\alpha}(T\varphi) = T(J^{-\alpha}\varphi) \in L^r$ so $T\varphi \in L^r_\alpha$.

(c) If $\alpha > n/r$, $\varphi \in \mathscr{S}$ then $T\varphi$ is equal to a continuous function h and $|h(0)| \le B\|\varphi\|_{p,\alpha}$ where B depends only on T, α, p and r:

From (b) $T\varphi \in L^r_\alpha$. From (10.3) $T\varphi$ is a continuous function h and $|h(0)| \leq A\|T\varphi\|_{r\alpha} = A\|J^{-\alpha}T\varphi\|_r = A\|T(J^{-\alpha}\varphi\|_r \leq A'\|J^{-\alpha}\varphi\|_p = A'\|\varphi\|_{p,\alpha}$.

(d) The mapping $\varphi \rightarrow h(0)$ is linear: Obvious.

(e) The mapping $\varphi \rightarrow h(0)$ is in $\mathcal{S}'$: In the proof of (9.1) we mentioned that if $\{\varphi_k\}$ is null in $\mathcal{S}$ then it is null in L^p_α. This shows that $\varphi \rightarrow h(0)$ is in $\mathcal{S}'$ since $|h(0)| \leq B\|\varphi\|_{p\alpha}$.

(f) Let $v \in \mathcal{S}'$ be such that $(v,\varphi) = h(0) = T\varphi\,(0)$. Then $T\varphi = \tilde{\tilde{v}} * \varphi$: $(\tilde{\tilde{v}} * \varphi)(x) = (\tilde{\tilde{v}}, \tau_x \tilde{\tilde{\varphi}}) = (\tilde{\tilde{v}}, (\tau_{-x}\varphi)^{\approx}) = (v, \tau_{-x}\varphi)$

$= (T(\tau_{-x}\varphi))(0) = (\tau_{-x}(T\varphi))(0) = (T\varphi)(x)$.

(g) If $w \in \mathcal{S}'$ is such that $w * \varphi = T\varphi$ for all $\varphi \in \mathcal{S}$ then $w = \tilde{\tilde{v}}$: Since $T\varphi = \tilde{\tilde{v}} * \varphi$, $\omega * \varphi = T\varphi$ implies, $(w - \tilde{\tilde{v}}) * \varphi = 0$ for all $\varphi \in \mathcal{S}$. That is $(w - \tilde{\tilde{v}}, \tau_x \tilde{\tilde{\varphi}})$ for all x and $\tilde{\tilde{\varphi}}$. Set $x = 0$. Then $(w - \tilde{\tilde{v}}, \tilde{\tilde{\varphi}}) = 0$ for all $\varphi \in \mathcal{S}$, so $w = \tilde{\tilde{v}}$ in $\mathcal{S}'$.

Let us briefly examine what sorts of distributions may appear in the relations $\varphi \rightarrow u * \varphi$ that is bounded from $L^p \cap \mathcal{S}$ into L^r for $u \in \mathcal{S}'$. The collection of such distributions is denoted T^r_p. If $u \in T^r_p$ then $T^r_p(u)$ denotes the norm of $\varphi \rightarrow u * \varphi$.

If $u \in L^1$ then $u \in T^p_p$, $1 \leq p \leq \infty$, $T^p_p(u) \leq \|u\|_1$.

If $u \in L^p$ then $u \in T_1^p$, $1 \le p \le \infty$, $T_1^p(u) \le \|u\|_p$, and $u \in T_{p'}^\infty$, $T_{p'}^\infty(u) = T_1^p$ as follows from (7.5). More generally (7.5) shows that $T_p^r = T_{r'}^{p'}$. Young's theorem on convolution says that if $u \in L^p$, $1 \le r \le p'$ and $1/s = (1/p) + (1/r) - 1$ then $u \in T_r^s$.

There are very few cases when we know the exact contents of T_r^s. One such case is $T_1^1 = T_\infty^\infty$. In this case $\Phi \longrightarrow (u * \varphi)(0)$ is a bounded linear functional on $\mathscr{S} \cap C_0$ and hence extends to a bounded linear functional on C_0, and this implies that $u \in M$ and $\|u\|_M = T_1^1(u) = T_\infty^\infty(u)$. Since u a finite Borel measure implies that $u \in T_1^1$, we see that $T_1^1 = T_\infty^\infty = M$ and $\|\cdot\|_M = T_1^1(\cdot) = T_\infty^\infty(\cdot)$.

One aspect of all the examples given is that when $u \in T_p^r$ we found $r \ge p$. This is a general property.

<u>Proposition (10.5)</u>. <u>If</u> $u \in T_p^r$ <u>and</u> $1 \le r < p \le \infty$ <u>then</u> $u = 0$.

<u>Proof</u>. If $\varphi \in \mathscr{S}$, $\|\varphi - \tau_h\varphi\|_p = 2^{1/p}\|\varphi\|_p$ if $|h|$ is large enough. (If φ is supported on $\mathfrak{B}^k$ then $h \notin \mathfrak{B}^k$ will do.) Since $T\varphi \in L^r$, $1 \le r < \infty$ we see that $\|T\varphi - \tau_h T\varphi\|_r \to 2^{1/r}\|T\varphi\|_r$ as $|h| \longrightarrow \infty$. (Approximate $T\varphi$ in L^r by functions in $\mathscr{S}$.)

Let A be the norm of T. Suppose $A > 0$. Then

$$\|T\varphi - \tau_h T\varphi\|_r = \|T\varphi - T(\tau_h\varphi)\|_r = \|T(\varphi - \tau_h\varphi)\|_r$$
$$\le A\|\varphi - \tau_h\varphi\|_p \to A\, 2^{1/p}\|\varphi\|_p .$$

Thus,

$2^{1/r}\|T\varphi\|_r \leq A\, 2^{1/p}\|\varphi\|_r$. Hence $\|T\varphi\|_r \leq A\, 2^{(1/p-1/r)}\|\varphi\|_r$, and $A\, 2^{(1/p-1/r)} < A$ if $A > 0$. This is a contradiction so $A = 0$. That is, $T\varphi = u * \varphi = 0$ for all $\varphi \in \mathscr{S}$ and this implies that $u = 0$.

Thus, $T_1^1 = T_\infty^\infty = M$, $T_p^r = \{0\}$ if $p > r$. The one other case that we can characterize simply is T_2^2.

<u>Theorem (10.6)</u>. $u \in T_2^2$ <u>iff</u> $\hat{u} \in L^\infty$. <u>If</u> $u \in L^\infty$ <u>then</u> $\|u\|_\infty = T_2^2(u)$.

<u>Proof.</u> If $\hat{u} \in L^\infty$ then for all $\varphi \in \mathscr{S}$, $\|u * \varphi\|_2 = \|(u * \varphi)^\wedge\|_2 = \|\hat{u}\hat{\varphi}\|_2 \leq \|\hat{u}\|_\infty \|\hat{\varphi}\|_2 = \|\hat{u}\|_\infty \|\varphi\|_2$. Thus, $u \in T_2^2$, $T_2^2(u) \leq \|\hat{u}\|_\infty$. In fact by the usual approximation argument $T_2^2(u) = \|\hat{u}\|_\infty$. Indeed, if $\hat{u}$ is of function type and $u \in T_2^2$, since we would have $\|\hat{u}\hat{\varphi}\|_2 < T_2^2(u)\|\hat{\varphi}\|_2^2$ for all $\varphi \in \mathscr{S}$, it is clear that $\hat{u} \in L^\infty$. Thus, to prove that $u \in T_2^2$ implies $\hat{u} \in L^\infty$ it will suffice to show that $\hat{u}$ is of function type.

Since $R(x,k) \in \mathscr{S}$ for all k we see that $u(x,k) \in L^2$ and so $\hat{u}\Phi_k \in L^2$. That is, $\hat{u}\Phi_k$ is a function. Let b be a measurable function defined by $b\Phi_k = \hat{u}\Phi_k$. b is well defined since $(\hat{u}\Phi_k)\Phi_\ell = \hat{u}\Phi_{k\vee\ell}$. We wish to show that $(u * \varphi)^\wedge = b\hat{\varphi}$ for all

$\varphi \in \mathcal{S}$. We may assume that $\hat{\varphi} = \hat{\varphi}\,\Phi_k$. Then $b\Phi_k = \hat{u}\,\Phi_k$ and

$$(u * \varphi)^{\wedge} = \hat{u}\,\hat{\varphi} = \hat{u}(\Phi_k\hat{\varphi}) = (\hat{u}\Phi_k)\hat{\varphi} = (b\,\Phi_k)\,\hat{\varphi} = b(\Phi_k\hat{\varphi}) = b\,\hat{\varphi}\,.$$

An Observation. *If* $p = 1$ or ∞, $\mathrm{Re}(\alpha) = 0$, $\alpha \neq 0$ *then* $L^p_\alpha \not\subset L^p$.

Proof. The claim is that J^α does not map L^p into L^p. If it did it would be bounded, as follows from the closed graph theorem. For let $\{(f_k, J^\alpha f_k)\}$ be a sequence in $L^p \times L^p$ and suppose $\{(f_k, J^\alpha f_k)\} \to (g,h)$, g and h in L^p. Then $\{f_k\} \to g$ in L^p and hence in $\mathcal{S}'$, and so $J^\alpha f_k \to J^\alpha g$ in $\mathcal{S}'$. But, $J^\alpha f_k \to h$ in L^p so $J^\alpha f_k \to h$ in $\mathcal{S}'$ and $J^\alpha g = h \in L^p$ and the graph is closed, so J^α is bounded. But then $G^\alpha \in T^p_p$ so G^α is a finite Borel measure. Since $R(x,k) \in C_0 \cap L^1$ it follows that $\|G^\alpha(\cdot,k)\|_1 = O(1)$ as $k \to -\infty$. But from (7.10) we get

$$\|G^\alpha(\cdot,k)\|_1 \geq (1/|\Gamma_n(\alpha)|) \int_{q^k \leq |x| \leq 1} |x|^{-n}dx - (1/|\Gamma_n(\alpha)|)\,q^{-n}(1-q^{kn})$$

$$= -k(1-q^{-n})/|\Gamma_n(\alpha)| + O(1), \quad \text{as} \quad k \to -\infty\,.$$

That is, $\|G^\alpha(\cdot,k)\|_1 \to +\infty$ as $k \to \infty$, thus $G^\alpha \notin M$ and $L^p_\alpha \not\subset L^p$.

11. Notes for Chapter III

A general reference for this chapter is Taibleson [3].

The notation we have used for "spheres" ($\mathfrak{P}^k, \mathfrak{D}, \mathfrak{P}^*$) needs some justification. We might have used a more normal looking notation such as S_k or B_k for a sphere of radius q^k. However, because we wished to emphasize the unity of the one- and n-dimensional cases and since the algebraic notation is so natural for the one-dimensional case we kept that notation.

§3. This section is concerned with two problems. The first is to define convolution and formulate the "equivalence" theorem, (3.15). The second is to identify, within the class of distributions $\mathcal{S}'$, a suitable class of multipliers, $\mathcal{O}_M$, on $\mathcal{S}'$; and a class of convolution distributions $\mathcal{O}_C$, on $\mathcal{S}'$. These identifications lead to the "big" theorem, (3.19).

It will be useful to keep in mind that the subscript "C" in "$\mathcal{O}_C$" is a mnemonic device to call "convolution" to mind, and it is a mere alphabetical coincidence that $\mathcal{O}_C$ is the class of distributions with compact support.

In the euclidean setting $\mathcal{O}_M$ is the class of infinitely differentiable functions all of whose derivatives are slowly increasing. A function, is said to be slowly increasing if for some positive integer k, $f(x)(1+|x|^2)^{-k}$ is bounded. For details see Schwartz [1]. A brief summary of what one needs to know about the euclidean case in order to be properly impressed with the analogous development for local fields can be found in Taibleson [6,pp.435-436].

§4. The main result of this section is (4.19) which we refer to (as is usual) as the theorem of Soboleff, but which might just as legitimately be called the theorem of Hardy and Littlewood-Soboleff since the one-dimensional version was established by Hardy and Littlewood. See Stein [2,p.165] for more of the background. The proof in these notes is that of Muckenhaupt and Stein as given in Stein [2]. For an Orlicz space version of these results see O'Neil [1].

We will need several results on the interpolation of operators. Most of what we need can be found in Stein and Weiss [2,Ch.V,pp.177-205] or in Zygmund [1,Ch.VII,§1,§4]. When we refer to the Riesz-Thorin interpolation theorem we will be referring to (1.11) in Zygmund, and when we refer to the Marcinkiewicz interpolation theorem we will be referring to (4.6) of Zygmund and we will use the estimates of the norms of the interpolated operators, (1.14), (4.7) and (4.21) without further specification.

§5. A treatment of Bessel potentials in the euclidean case can be found in Aronszajn and Smith [1]. A proof of (5.7) for the euclidean case is in Taibleson [6,Ch.IV,p.428].

§6. The results of this section for the euclidean case were first announced by Stein [4].

§7. The euclidean version of the results in this section is found in Calderón [3].

We note that (7.6) is due to Phillips [1] and is the local field version of the Calderón-Zygmund decomposition which is found in its original form in Calderón and Zygmund [1].

The function $G^{\alpha}(x,k)$ in (7.10) is defined by: $G^{\alpha}(\cdot,k) = G^{\alpha} * R(\cdot,k$ as is implicit in the statement of (4.11). Also, see the definition of a regularization preceding (1.5) in Chapter IV, below.

The cases left open by (7.14) are $Re(\alpha-\beta) = n(1/p-1/r)$, with $p =$ or $r = \infty$. For simplicity we consider only the case $\beta = 0$.

If $p = r$ then either $p = 1$ or $p = \infty$ and $Re(\alpha) = 0$. That case is treated in §10. The result is that if $\alpha = 0$ the inclusion result holds (trivially) and if $\alpha = 0$ the result fails. (The ambitious reader can check this by examining the representations of G^{α} in (5.2) for $Re(\alpha) =$

If $\text{Re}(\alpha) > 0$, then we have several possible cases. If $1 = p < r < \infty$, $\text{Re}(\alpha) = n/r'$ and the inclusion fails, although we do get the weak-type result that is implicit in the proof of (4.9). Similarly if $1 < p < r = \infty$, $\text{Re}(\alpha) = n/p$, the inclusion also fails. If $p = 1$, $r = \infty$, $\text{Re}(\alpha) = n$, then the validity of the result depends on whether or not $\alpha = n$ or $\alpha \neq n$. If $\alpha \neq n$ we see from (5.2) that $G^\alpha \in L^\infty$ so $L^1_\alpha \subset L^\infty_0$, and the inclusion is continuous. However, if $\alpha = n$, and we use the description of G^n in (5.2), it is easy to construct an $f \in L^1$ such that $G^n * f \notin L^\infty$ so $L^1_n \not\subset L^\infty_0$.

§9. For background on this section see Calderón [3].

§10. A general background for this section can be found in Hörmander [1]. Especial attention is called to (10.4), and it is suggested that proof of the analogous result in Hörmander be compared to that in the notes.

We also suggest that the reader compare the proof of (10.6) with that of the corresponding result, Theorem 3.18 in Stein and Weiss [2,Ch.I,p.28].

Chapter IV. Regularization and the theory of regular and sub-regular functions

In §1 of this chapter we systematically investigate the regularization kernels $R(x,k)$ and the regularizations $f(x,k)$ for $f \in \mathcal{S}'$. We study the kernels $R(x,k)$ as analogues of the Poisson kernels and the regularizations $f(x,k)$ as analogues of harmonic functions. After a brief look (in §2) at Lipschitz theory we turn in §3 to notions of sub-regular functions, least regular majorants and of non-tangential convergence.

1. Regular functions on $K^n \times \mathbb{Z}$

Consider the Poisson kernel $P(x,y)$, $x \in R^n$, $y > 0$. The properties of $P(x,y)$ that are of crucial interest are:

(a) $P(x,y) = y^{-n} P(xy^{-1},1)$, $y > 0$; (b) $P(x,y) \geq 0$; (c) $\int P(x,y)dx = 1$, $y > 0$; (d) $P(x,y)$ is radial in x; (e) $P(\cdot,y_1) * P(\cdot,y_2) = P(\cdot,y_1 + y_2)$

(f) $P(x,y)$ satisfies a nice differential equation (Laplace's).

The "central" property is the semi-group property expressed in (e). It says that convolution is functorial with respect to addition in the valuation group, $(0,\infty)$.

The valuation group of K^n can be identified with $\mathbb{Z}(\{p^k\}_{k\in\mathbb{Z}})$. The analogous family of functions would be functions $R(x,k)$ defined on $K^n \times \mathbb{Z}$ for which

(a′) $R(x,k) = |p^k|^n R(xp^k,0)$, $\forall\, k \in \mathbb{Z}$; (b′) $R(x,k) \geq 0$;

(c′) $\int R(x,k)dx = 1$, $k \in \mathbb{Z}$; (d′) $R(x,k)$ is radial in x;

(e′) $R(\cdot,k_1) * R(\cdot,k_2) = R(\cdot,k_1 \vee k_2)$; (f′) $R(x,k)$ satisfies a "nice" difference equation.

We will now show that (up to a scale factor) there is exactly one family of functions which satisfy (a′) - (e′).

Let $f(x) = R(x,0)$. From (e′) with $k_1 = k_2 = 0$ we get $f * f = f$. (b′) and (c′) show that $f \in L^1$ so $(\hat{f})^2 = \hat{f}$. Thus, $\hat{f}(\xi)$ is equal to zero or one and hence is the characteristic function of a set. Since f is radial, (d′), we see that $\hat{f}$ is radial, and so the set is radial. Now note that $\hat{f}(\xi) \to 0$ as $|\xi| \to \infty$ and $\hat{f}$ is continuous. This implies that the set has compact support. Since $\hat{f}(0) = \int f(x)dx = 1$ and $\hat{f}$ is continuous, the set contains a ball about the origin. We show, it must be the characteristic function of a ball. We know it is a radial set with compact support that contains a ball. From (a′) and (e′) we see that if $k \leq 0$, then $\hat{f}(\xi p^{-k})\hat{f}(\xi) = \hat{f}(\xi)$. This shows that if $\hat{f}(\xi) \neq 0$, then $\hat{f}(\xi p^{-k}) \neq 0$ for all $k \leq 0$. Thus $\hat{R}(\xi,0) = \hat{f}(\xi) = \Phi_k(\xi)$ for some k.

We scale by choosing $k = 0$, and it follows that

$$R(x,k) = q^{-kn}\Phi_{-k}(x), \quad \hat{R}(\xi,k) = \Phi_k(\xi).$$

We now examine the difference equation satisfied by $R(x,k)$. Note that, as a function of x, $R(x,k)$ is constant on cosets of $\mathfrak{P}^{-k}$. Let $\{a_s^k\}_{s=1}^{q^n}$ be a complete set of coset representatives of $\mathfrak{P}^{-k}$ in $\mathfrak{P}^{-(k+1)}$. Then

$$(1.1) \qquad R(x,k+1) = q^{-n}\sum_{s=1}^{q^n} R(x+a_s^k, k), \; x \in K^n, \; k \in \mathbb{Z}.$$

Equivalently,

$$(1.2) \qquad \int_{y+\mathfrak{P}^{-\ell}} R(x,k)\,dx = \int_{y+\mathfrak{P}^{-\ell}} R(x,\ell)\,dx, \; y \in K^n, \; k \le \ell .$$

These facts are established by inspection.

Definition. We say that $f(x,k)$ defined on $K^n \times \mathbb{Z}$ is _regular_ if:

(1.3)

(i) $f(x,k)$ is constant on cosets of $\mathfrak{P}^{-k}$ as a function of x;

(ii) $f(x,k+1) = q^{-n}\sum_{s=1}^{q^n} f(x+a_s^k,k), \; x \in K^n, \; k \in \mathbb{Z}$

Equivalent to (ii) is

(ii′) $\displaystyle\int_{y+\mathfrak{P}^{-\ell}} f(x,k)\,dx = \int_{y+\mathfrak{P}^{-\ell}} f(x,\ell)\,dx, \; y \in K^n, \; k \le \ell$

<u>Proposition (1.4)</u>. <u>If</u> $f(x,k)$ <u>is regular</u>, $k_1, k_2 \in \mathbb{Z}$ <u>then</u>

$f(\cdot,k_1) * R(\cdot,k_2) = f(\cdot,k_1 \vee k_2)$.

<u>Proof</u>. $f(x,k)$ is locally bounded and $R(x,k)$ is bounded with compact support so the integral exists.

$$f(\cdot,k_1) * R(\cdot,k_2)(x) = \int f(z,k_1)R(x-z,k_2)dz$$

$$= q^{-k_2} \int_{z \in x + \mathfrak{P}^{-k_2}} f(z,k_1)dz \ .$$

If $k_1 \geq k_2$ then $f(z,k_1)$ is constant on $x + \mathfrak{P}^{-k_2}$ so

$f(\cdot,k_1) * R(\cdot,k_2) = f(\cdot,k_1) = f(\cdot,k_1 \vee k_2)$ (by (1.3)(i)).

If $k_2 \geq k_1$, then (1.3)(ii) says that we have

$$f(\cdot,k_1) * R(\cdot,k_2)(x) = q^{-k_2} \int_{z \in x + \mathfrak{P}^{-k_2}} f(z,k_2)dz = f(x,k_2) = f(x,k_1 \vee k_2).$$

We recall our definition of $f(x,k)$ for $f \in \mathcal{S}'$ (III (4.11)).

<u>Definition.</u> If $f \in \mathcal{S}'$, $f(x,k) = (f * R(\cdot,k))(x)$, so $\hat{f}(\cdot,k) = \hat{f}\Phi_k$.

$f(x,k)$ is called the <u>regularization</u> of f.

<u>Lemma (1.5)</u>. <u>The regularization of a distribution</u> $f \in \mathcal{S}'(K^n)$ <u>is a regular function on</u> $K^n \times \mathbb{Z}$ <u>which converges to</u> f <u>in the</u>

<u>topology of $\mathcal{S}'$ as $k \to -\infty$. Conversely, every regular function is the regularization of some distribution.</u>

<u>Proof</u>. $f(x,k) = (f, \tau_x R(\cdot,k))$. Since $R(\cdot,k)$ is constant on cosets of $\mathfrak{P}^{-k}$, so if $f(\cdot,k)$. Now let $\{a_s^k\}_{s=1}^{q^n}$ be as in (1.1) - (1.3). Then $q^{-n}\sum_{s=1}^{q^n} f(x + a_s^k, k) = q^{-n}\sum_{s=1}^{q^n} (f, \tau_{x+a_s^k} R(\cdot,k))$

$= (f, q^{-n}\sum_s \tau_{x+a_s^k} R(\cdot,k)) = (f, \tau_x R(\cdot,k+1)) = f(x,k+1)$. Thus, $f(x,k)$ is regular. To see that $f(\cdot,k) \to f$ in $\mathcal{S}'$ as $k \to -\infty$, fix $\varphi \in \mathcal{S}$. $\varphi^\vee$ is supported on some $\mathfrak{P}^{-\ell}$. If $k \le \ell$ then $\Phi_k\varphi^\vee = \varphi^\vee$ so $(f(\cdot,k),\varphi) = (\hat{f}(\cdot,k),\varphi^\vee) = (\hat{f}\Phi_k, \varphi^\vee) = (\hat{f},\varphi^\vee) = (f,\varphi)$, when $k \le \ell$.

Assume now that $u(x,k)$ is regular. Then $(u(\cdot,k),\varphi) = \int u(x,k)\varphi(x)dx$ is defined. Let φ be constant on cosets of $\mathfrak{P}^s$ and write $\varphi(x) = \sum a_\ell \tau_{h_\ell}\Phi_s(x)$. Then

$$(u(\cdot,k),\varphi) = \sum_\ell a_\ell \int_{h_\ell+\mathfrak{P}^s} u(x,k)dx = \sum_\ell a_\ell \int_{h_\ell+\mathfrak{P}^s} u(x,-s)dx, \quad \text{if } k \le -s.$$

Thus, if we let $(f, \tau_h \Phi_s) = \int_{h+\mathfrak{P}^s} u(x,-s)dx$, it is easy to check that

$f \in \mathcal{S}'$. It is also immediate that $u(x,k) \to f$ in $\mathcal{S}'$ and $f(x,k) = u(x,k)$. For this last step,

$$f(x,k) = q^{-kn}(f, \tau_x \Phi_{-k}) = q^{-kn}\int_{x+\mathfrak{P}^{-k}} u(z,k)dz = u(x,k).$$

<u>Corollary (1.6)</u>. <u>If</u> $f(x,k)$ <u>is regular and</u> $f(x,k) \geq 0$ <u>for all</u> $(x,k) \in K^n \times \mathbb{Z}$ <u>then it is the regularization of a Borel measure on</u> K^n.

<u>Proof</u>. Let f be the distribution to which $f(x,k)$ converges. Take $\varphi \in \mathcal{S}$ and suppose it is supported on $\mathfrak{P}^s$. For k small enough, $|(f,\varphi)| = |\int_{\mathfrak{P}^s} f(x,k)\varphi(x)dx| \leq \|\varphi\|_\infty \int_{\mathfrak{P}^s} f(x,k)dx = \|\varphi\|_\infty \int_{\mathfrak{P}^s} f(x,-s)dx$

$= \|\varphi\|_\infty q^{-ns} f(0,s)$. An application of the Riesz representation theorem shows that $f\Phi_s$ is a finite Borel measure and so f is a Borel measure.

Recall the definition of the Hardy-Littlewood maximal function. In the current notation $Mf(x) = \sup_k |f|(x,k) = \sup_k q^{-kn} \int_{u \in \mathfrak{P}^{-k}} |f(x+u)| du$.

<u>Theorem (1.7)</u>. <u>If</u> $f \in L^p(K^n)$, $1 < p \leq \infty$, <u>then</u> $Mf \in L^p$ <u>and there is a constant</u> $A_p > 0$ <u>independent of</u> f <u>such that</u> $\|Mf\|_p \leq A_p \|f\|_p$.

<u>Proof</u>. (i) $f \to Mf$ is sub-linear. (ii) $f \in L^\infty$ implies $\|Mf\|_\infty \leq \|f\|_\infty$. Obvious. (iii) Using III (1.13), if $f \in L^1$, $|\{x : Mf(x) > s\}| \leq \|f\|_1 s^{-1}$ for all $s > 0$. The result now follows from the Marcinkiewicz interpolation theorem.

<u>Remark</u>. Recall that if f is locally integrable then $f(x,k) \to f(x)$ a.e. III (1.14).

<u>Theorem (1.8)</u>. (a) $\|f(\cdot,k)\|_p \leq \|f\|_p$, $k \in \mathbb{Z}$

(b) $f(x,k) \to f(x)$ a.e. as $k \to -\infty$

(c) $\|f(\cdot,k)\|_p \to \|f\|_p$ as $k \to -\infty$

$\Big\}$ $f \in L^p$, $1 \leq p \leq \infty$

(d) $f(\cdot,k) \to f$ in L^p, $1 \leq p < \infty$

$f(\cdot,k) \to f$ in the Weak* topology, $p = \infty$

$\Big\}$ as $k \to -\infty$, $f \in L^p$.

(e) <u>If</u> $f \in M$; $\|f(\cdot,k)\|_1 \leq \|f\|_M$, $k \in \mathbb{Z}$; $\|f(\cdot,k)\|_1 \to \|f\|_M$, $k \to -\infty$; <u>and</u> $f(\cdot,k) \to f$ <u>in the Weak* topology as</u> $k \to -\infty$

(f) <u>If</u> f <u>is uniformly continuous and bounded then</u> $f(x,k) \to f(x)$ <u>uniformly as</u> $k \to -\infty$.

<u>Proof</u>. Since $f \in L^p$ implies f is locally integrable (b) follows from III (1.14).

The function $R(\cdot,k)$ satisfies $\|R(\cdot,k)\|_1 = 1$ for all k, so if $f \in L^p$, $f(\cdot,k) = f * R(\cdot,k) \in L^p$ and the map $f \to f(\cdot,k)$ is bounded with norm 1 from L^p to L^p. Similarly from M to L^1. This is (a) and the first part of (e).

In III (4.11) we established (f) and the first part of (d). For the last part of (d) we view L^∞ as $(L^1)^*$. We wish to show that if $f \in L^\infty$, $g \in L^1$ then $\int f(x,k)g(x) \to \int f(x)g(x)$ as $k \to -\infty$. But an easy application of Fubini's theorem gives $\int f(x,k)g(x)dx = \int f(x)g(x,k)dx$. It follows from (d) with $p = 1$ that $\int f(x)g(x,k)dx \to \int f(x)g(x)dx$. That completes the proof of (d).

The last part of (e) follows by a similar argument, using (f) applied to $g \in C_0$.

For $1 \leq p < \infty$, (c) follows immediately from (d). If $f \in L^\infty$, and $\epsilon > 0$, there is a $g \in L^1$ such that $\|g\|_1 = 1$ and $\int fg > \|f\|_\infty - \epsilon$. For all $k \in \mathbb{Z}$, $\|f(\cdot,k)\|_\infty \geq |\int f(x,k)g(x)dx|$ $\geq |\int f(x)g(x)dx| - |\int(f(x,k)-f(x))g(x)dx|$.

Since the first term dominates $\|f\|_\infty - \epsilon$ and the second is $o(1)$ as $k \to -\infty$ we see that $\liminf_k \|f(\cdot,k)\|_\infty \geq \|f\|_\infty$. But $\|f(\cdot,k)\|_\infty \leq \|f\|_\infty$, so $\lim_k \|f(\cdot,k)\|_\infty = \|f\|_\infty$. This completes the proof of (c). The middle part of (e) is established by an analogous argument that is left as an exercise. This completes the proof.

We now supply a "converse" to (1.8)(a). For $1 < p \leq \infty$ the converse is complete.

Theorem (1.9). *Suppose* $f(x,k)$ *is regular and for some fixed* p, $1 \leq p \leq \infty$, $\|f(\cdot,k)\|_p \leq A$ *for some* $A > 0$ *independent of* $k \in \mathbb{Z}$. *Then:*

(a) *If* $1 < p \leq \infty$, $f(x,k)$ *is the regularization of a function in* L^p.

(b) *If* $p = 1$, $f(x,k)$ *is the regularization of a finite Borel measure*.

(c) *If* $p = 1$ *and* $\{f(\cdot,k)\}$ *is Cauchy in* L^1 *as* $k \to -\infty$ *then* $f(x,k)$ *is the regularization of a function in* L^1.

(d) <u>If</u> $p = \infty$ <u>and</u> $\{f(\cdot,k)\}$ <u>is Cauchy in</u> L^∞ <u>as</u> $k \to -\infty$ <u>then</u> $f(x,k)$ <u>is the regularization of a bounded uniformly continuous functio</u>

<u>Proof</u>. Suppose $1 < p \leq \infty$, $\|f(\cdot,k)\|_p \leq A$ for all k, and $f(x,k)$ is regular. Let r be conjugate to p. That is, $(1/r) + (1/p) = 1$. Then there is a sequence $\{k_t\}, k_t \to -\infty$ and a function $F \in L^p$ such that $f(\cdot,k_t) \to F$ in the Weak* topology. That is, for any $g \in L^r$, $\int f(x,k_t)g(x)dx \to \int F(x)g(x)dx$. Then $\int f(z,k_t)R(x-z,k)dz \to \int F(z)R(x-z,k)dz = F(x,k)$. But for $k_t \geq k$, the left hand side is $f(x,k)$, (1.4), so $f(x,k) = F(x,k)$, which proves (a).

For (b) note that $R(\cdot,k) \in C_0$ and that L^1 is embedded isometrically in M in the usual way. Then (b) follows by the identical argument.

For (d) we note that $f(x,k)$ is uniformly continuous in x for each k, and is bounded uniformly in x and k, and converges uniformly. It is now a standard exercise that $f(x,k)$ converges uniformly to a bounded uniformly continuous function. From (a) we have that $f(x,k)$ is the regularization of a bounded function F to which is converges a.e. ((1.8)(b)). This implies that F is the uniformly continuous function.

For (d) we use an obvious adaptation of the argument for (d).

<u>Notation</u>. p' is the conjugate index to p, $1 \leq p \leq \infty$. $(1/p)+(1/p')=1$, so $p' = p/(p-1)$. We use the convention $1/\infty = 0$ so if $p = 1$, $p' = \infty$, and if $p = \infty$, $p' = 1$.

<u>Lemma (1.10)</u>. (a) $\|R(\cdot,k)\|_p \leq q^{-nk/p'}$, $1 \leq p \leq \infty$.

(b) <u>If</u> $f \in L^p(K^n)$, $1 \leq p \leq \infty$, <u>then</u> $\|f(\cdot,k)\|_p \leq \|f\|_\infty q^{-nk/p}$.

<u>Proof</u>. (a) follows by simple computation. To obtain (b) we have $\|f(\cdot,k)\|_p \leq \|f\|_p \|R(\cdot,k)\|_{p'}$ and use (a).

<u>Lemma (1.11)</u>. <u>Let</u> p <u>and</u> r <u>be given</u>, $1 \leq p, r \leq \infty$, $(1/p)+(1/r) \geq 1$, $k_1,k_2 \in \mathbb{Z}$. <u>Let</u> $f(x,k), g(x,k)$ <u>be regular functions such that</u> $f(\cdot,k) \in L^p$, $g(\cdot,k) \in L^r$ <u>for all</u> k. <u>Then</u>:

(a) $\int f(x,k_1)g(x,k_2)dx = \int f(x,k_1 \vee k_2)g(x,k_1 \vee k_2)dx$

(b) $f(\cdot,k_1) * g(\cdot,k_2) = h(\cdot,k_1 \vee k_2)$ <u>is a well defined regular function</u>.

<u>Proof</u>. Since $f(\cdot,k)$ is regular and in L^p, then since $f(\cdot,k) = f(\cdot,k) * R(\cdot,k)$ it follows that $f(\cdot,k) \in L^\infty$ so $f(\cdot,k) \in L^s$, $p \leq s \leq \infty$. Thus, $f(\cdot,k) \in L^{r'}$ so $\int f(x,k_1)g(x,k_2)dx$ converges absolutely. Suppose $k_2 \geq k_1$. Then Fubini's theorem and the fact that $R(\cdot,k)$ is even gives us,

$$\int f(x,k_1)g(x,k_2)dx = \int f(x,k_1)(g(\circ,k_2) * R(\cdot,k_2))(x)dx$$
$$= \int g(x,k_2)(f(\cdot,k_1) * R(\cdot,k_2))(x)dx = \int g(x,k_2)f(x,k_2)dx .$$

This proves (a) and shows that the convolution in (b) is well defined for all x and k, and we may define $h(\cdot,k) = f(\cdot,k) * g(\cdot,k)$. Since $\tau_u h(\cdot,k) = (\tau_u f(\cdot,k)) * g(\cdot,k)$ we see that h is constant on cosets of $\mathfrak{P}^{-k}$ since f is. Now using the notation of (1.3) we have $q^{-n} \sum_s \tau_{a_s^k} h(\cdot,k) = (q^{-n} \sum_s \tau_{a_s^k} f(\cdot,k)) * g(\cdot,k) = f(\cdot,k+1) * g(\cdot,k)$

$= h(\cdot,(k+1) \vee k) = h(\circ,k+1)$. This shows that h is regular.

The fact that $\lim_{k \to -\infty} \|f(\cdot,k)\|_\infty = \|f\|_\infty$ when $f \in L^\infty$ can be restated as a maximal principle. Note that $f(\cdot,k+1) = f(\cdot,k) * R(\cdot,k+1$ so $\|f(\cdot,k+1)\|_\infty \leq \|f(\cdot,k)\|_\infty$ so the sequence $\{\|f(\circ,k)\|_\infty\}$ increases as $k \to -\infty$. Thus, $\sup_{(x,k) \in K^n \times \mathbb{Z}} |f(x,k)| = \|f\|_\infty$, and so:

<u>Proposition (1.12)</u>. (Maximal Principle). <u>Let</u> $f(x,k)$ <u>be a bounded regular function on</u> $K^n \times \mathbb{Z}$ <u>and let</u> F <u>be a function to which</u> $f(x,k)$ <u>converges a.e. as</u> $k \to -\infty$. <u>Then</u> $F \in L^\infty$ <u>and</u>

$$\operatorname*{ess\,sup}_{x \in K^n} |F(x)| = \sup_{(x,k) \in K^n \times \mathbb{Z}} |f(x,k)| .$$

<u>Proof</u>. The conditions imply that $f(x,k)$ is regularization of F. The rest is immediate.

<u>Corollary (1.13)</u>. <u>Let</u> $f(x,k)$ <u>be a bounded regular function on</u> $K^n \times \mathbb{Z}$ <u>and suppose</u> $f(x,k) \to 0$ <u>for a.e.</u> x <u>as</u> $k \to -\infty$. <u>Then</u> $f(x,k) \equiv 0$ <u>on</u> $K^n \times \mathbb{Z}$.

2. More about Lipschitz conditions

The following result is the discrete analogue of Hardy's inequality.

<u>Lemma (2.1)</u>. <u>Suppose</u> $\alpha > 0$ <u>and</u> $\{a_k\}_{k\in\mathbb{Z}}$ <u>is a sequence of non-negative numbers such that</u>:

$$\left[\sum (q^{-k\alpha} a_k)^p\right]^{1/p} = \|\{q^{-k\alpha} a_k\}\|_p < \infty \ , \ 1 \le p < \infty \ , \ \text{or}$$

$$\sup_k q^{-k\alpha} a_k = \|\{q^{-k\alpha} a_k\}\|_\infty < \infty \ , \ p = \infty \ .$$

<u>Let</u> $A_k = \sum_{\ell=-\infty}^{k} a_\ell$. <u>Then</u>

$$\|\{q^{-k\alpha} A_k\}\|_p \le (1-q^{-\alpha})^{-1} \|\{q^{-k\alpha} a_k\}\|_p \ , \ 1 \le p \le \infty \ .$$

<u>Proof</u>. Let $b_k = q^{-k\alpha} a_k$, $B_k = q^{-k\alpha} A_k$. Then

$$B_k = q^{-k\alpha} \sum_{\ell=-\infty}^{k} a_\ell = \sum_{\ell=-\infty}^{k} q^{-(k-\ell)\alpha} b_\ell = \sum c_{k-\ell}\, b_\ell \ ,$$

where $c_k = \begin{cases} q^{-k\alpha} & , k \ge 0 \\ 0 & , k < 0 \end{cases}$. But $\{c_k\} \in \ell^1$,

$\|\{c_k\}\|_1 = \sum_{k=0}^{\infty} q^{-k\alpha} = (1-q^{-\alpha})^{-1}$, and $\{B_k\} = \{b_k\} * \{c_k\}$.

Thus, $\|\{q^{-k\alpha} A_k\}\|_p = \|\{B_k\}\|_p \leq \|\{c_k\}\|_1 \; \|\{b_k\}\|_p = (1-q^{-\alpha})^{-1} \|\{q^{-k\alpha} a_k\}\|_p$.

Notation. If $g(x,h)$ is a function on $K^n \times K^n$ we set,

$$\|g(x,h)\|_{pr} = \left[\int_{K^n} \|g(\cdot,h)\|_p^r \; |h|^{-n} dh \right]^{1/r}, \quad 1 \leq r < \infty$$

$$\|g(x,h)\|_{p\infty} = \text{ess sup}_{h \in K^n} \|g(\cdot,h)\|_p \quad .$$

If $g(x,k)$ is a function on $K^n \times \mathbb{Z}$ (not necessarily regular) we set,

$$\|g(x,k)\|_{pr} = \|\{\|g(\cdot,k)\|_p\}\|_r \quad , \quad 1 \leq r \leq \infty \quad .$$

Theorem (2.2). If $f \in L^p(K^n)$ for some p, $1 \leq p \leq \infty$, $\alpha > 0$ and $1 \leq s \leq \infty$, then the following "norms" are equivalent:

A: $\| \, |h|^{-\alpha} (f(x+h) - f(x)) \|_{ps}$

B: $\|q^{-k\alpha} (f(x,k) - f(x))\|_{ps}$

C: $\|q^{-k\alpha} (f(x,k) - f(x,k-1))\|_{ps}$,

in the sense that if any one of them is finite, then so are the others and their ratios are bounded above and below by positive constants.

<u>Proof</u>. We will show that $A \le 2B$; $B \le 2A$; $C \le (1+q^{-\alpha})B$; and $B \le (1-q^{-\alpha})^{-1}C$.

<u>$A \le 2B$:</u> If $|h| = q^k$ then $f(x,k) = f(x+h,k)$. Thus,

$f(x+h) - f(x) = (f(x+h) - f(x+h,k)) + (f(x,k) - f(x))$, and

$\|f(\cdot + h) - f(\cdot)\|_p \le 2\|f(\cdot,k) - f(\cdot)\|_p$ if $|h| = q^k$.

If $s = \infty$, then $A = \sup_{h\in K^n} |h|^{-\alpha}\|f(\cdot + h) - f(\cdot)\|_p$

$$= \sup_{\substack{k\in \mathbb{Z} \\ |h|=q^k}} q^{-k\alpha}\|f(\cdot + h) - f(\cdot)\|_p \le \sup_{k\in\mathbb{Z}} 2\, q^{-k\alpha}\|f(\cdot,k) - f(\cdot)\|_p = 2\,B.$$

If $1 \le s < \infty$, then

$$A = \left[\int_{h\in K^n} (|h|^{-\alpha}\|f(\cdot + h) - f(\cdot)\|_p)^s\, |h|^{-n} dh\right]^{1/s}$$

$$= \left[\sum_{k\in\mathbb{Z}} \int_{|h|=q^k} q^{-kn}(q^{-k\alpha}\|f(\cdot + h) - f(\cdot)\|_p)^s\, dh\right]^{1/s}$$

$$\le 2\left[\sum_{k\in\mathbb{Z}} (1-q^{-n})q^{-k\alpha}\|f(\cdot,k) - f(\cdot)\|_p)^s\right]^{1/s}$$

$$= 2(1-q^{-n})^{1/s}\left[\sum_{k\in\mathbb{Z}} (q^{-k\alpha}\|f(\cdot,k) - f(\cdot)\|_p)^s\right]^{1/s} = 2(1-q^{-n})^{1/s}\, B.$$

<u>$B \le 2A$:</u> $\quad f(x,k) - f(x) = \int (f(x-z) - f(x))R(z,k)\,dz,$

$$\|f(\cdot,k) - f(\cdot)\|_p \le q^{-kn} \int_{|z| \le q^k} \|f(\cdot + z) - f(\cdot)\|_p dz$$

$$= q^{-kn} \sum_{\ell=-\infty}^{k} \int_{|z|=q^\ell} \|f(\cdot + z) - f(\cdot)\|_p dz .$$

From (2.1) we obtain,

$$B \le (1-q^{-(\alpha+n)})^{-1} \| \{ q^{-k(\alpha+n)} \int_{|z|=q^k} \|f(\cdot + z) - f(\cdot)\|_p dz \} \|_s \quad .$$

Thus for $1 \le s < \infty$ we have,

$$B \le 2 \Big\{ \sum_{k \in \mathbb{Z}} \Big[q^{-k\alpha}(1-q^{-n}) \{ q^{-kn}(1-q^{-n})^{-1} \int_{|z|=q^k} \|f(\cdot + z) - f(\cdot)\|_p dz \} \Big]^s \Big\}^{1/s}$$

$$\le 2 \Big\{ \sum_{k \in \mathbb{Z}} \Big[q^{-k\alpha} (1-q^{-n}) \{ q^{-kn}(1-q^{-n})^{-1} \int_{|z| \le q^k} \|f(\cdot + z) - f(\cdot)\|_p^s dz \}^{1/s} \Big]^s \Big\}^{1/s}$$

$$= 2(1-q^{-n})^{1/s'} \Big\{ \sum_{k \in \mathbb{Z}} \int_{|z|=q^k} (|z|^{-\alpha} \|f(\cdot + z) - f(\cdot)\|_p)^s |z|^{-n} dz \Big\}^{1/s}$$

$$\le 2 \Big\{ \int_{K^n} (|z|^{-\alpha} \|f(\cdot + z) - f(\cdot)\|_p)^s |z|^{-n} dz \Big\}^{1/s} = 2A .$$

For $s = \infty$, an analogous, but easier argument works.

$\underline{C \le (1 + q^{-\alpha})B}$: $q^{-k\alpha} \| f(\cdot,k) - f(\cdot,k-1) \|_p$

$$\le q^{-k\alpha} \{ \| f(\cdot,k) - f(\cdot) \|_p + \| f(\cdot,k-1) - f(\cdot) \|_p \}$$

$$= q^{-k\alpha} \| f(\cdot,k) - f(\cdot) \|_p + q^{-\alpha} \cdot q^{-(k-1)\alpha} \| f(\cdot,k-1) - f(\cdot) \|_p .$$

The result is now immediate.

$\underline{B \le (1-q^{-\alpha})^{-1}}$: For a.e. x, $f(x,k) - f(x) = \sum_{\ell=-\infty}^{k} (f(x,\ell) - f(x,\ell-1))$.

In fact, the right hand side converges to the left hand side in L^p, $1 \le p < \infty$. Thus, $\| f(\cdot,k) - f(\cdot) \|_p \le \sum_{\ell=-\infty}^{k} \| f(\cdot,\ell) - f(\cdot,\ell-1) \|_p$.

The result now follows from (2.1).

3. Sub-regular functions, domains, regular majorants and non-tangential convergence.

We write $(x + \mathfrak{P}^{-k},\ell) = \{(y,\ell) \in K^n \times \mathbb{Z} : y \in x + \mathfrak{P}^{-k}\}$. If $\mathcal{D} \subset K^n \times \mathbb{Z}$ then $\partial \mathcal{D} = \{(x,k) \in \mathcal{D} : (x,k-1) \notin \mathcal{D}\}$. A set $\mathcal{D} \subset K^n \times \mathbb{Z}$ is called a <u>domain</u> in $K^n \times \mathbb{Z}$ if:

(i) $(x,k) \in \mathcal{D} \Longrightarrow (x + \mathfrak{P}^{-k},k) \subset \mathcal{D}$

(ii) $(x,k) \in \mathcal{D}$ and $(x,k-1) \in \mathcal{D} \Longrightarrow (x + \mathfrak{P}^{-k},k-1) \subset \mathcal{D}$

(iii) $(x,k) \in \partial \mathcal{D} \Longrightarrow (x,\ell) \notin \mathcal{D}\ \forall\ \ell < k$.

A domain $\mathcal{D} \subset K^n \times \mathbb{Z}$ is <u>bounded</u> if $\exists\, k_0 \in \mathbb{Z} \ni k \geq k_0$ for all $(x,k) \in \mathcal{D}$.

For $\mathcal{D} \subset K^n \times \mathbb{Z}$, a domain we define a function on K^n,

$$m(x) = \begin{cases} \sup\{k : (x,k) \in \mathcal{D}\} & \text{if } (x,k) \in \mathcal{D} \text{ for some } k\,. \\ +\infty\,, & \text{if } (x,k) \notin \mathcal{D} \text{ all } k. \end{cases}$$

We say that $\mathcal{D}$ is <u>simple</u> if $m(\mathcal{D}) = \inf_{x \in K^n} m(x) > -\infty$.

A <u>path</u> in a domain $\mathcal{D}$ is a collection $\{(x_s + \mathfrak{B}^{-k_s}, k_s)\}_{s=0}^{r}$ such that: (i) $(x_s + \mathfrak{B}^{-k_s}, k_s) \subset \mathcal{D}$, $s = 0,1,\ldots,r$;

(ii) $k_s = k_{s-1} \pm 1$, $s = 1,2,\ldots,r$; and

(iii) $(x_s + \mathfrak{B}^{-k_s}) \cap (x_{s-1} + \mathfrak{B}^{-k_{s-1}}) \neq \phi$.

A domain $\mathcal{D}$ is <u>connected</u> if for every pair (x,k), $(y,\ell) \in \mathcal{D}$ there is a path $\{(x_s + \mathfrak{B}^{-k_s}, k_s)\}_{s=0}^{r}$ in $\mathcal{D}$ that connects the pair in the sense that $(x_0,k_0) = (x,k)$ and $(y,\ell) = (x_r,k_r)$.

<u>Remark</u>. If $\mathcal{D}$ is a connected domain, then it is simple. This follows from the observation that if (x,k), $(y,\ell) \in \mathcal{D}$ and $\mathcal{D}$ is connected then $m(x) = m(y) \geq k \vee \ell$.

A function $f(x,k)$ on a domain $\mathcal{D}$ in $K^n \times \mathbb{Z}$ is <u>smooth</u> if $f(x,k)$ is constant on $(x + \mathfrak{B}^{-k}, k)$.

A function $f(x,k)$ on a domain $\mathcal{D}$ in $K^n \times \mathbb{Z}$ is regular if $f(x,k)$ is smooth and for all $(x,k) \in \mathcal{D} \approx \partial\mathcal{D}$,

$$(3.1) \qquad f(x,k) = q^{-kn} \int_{x+\mathfrak{B}^{-k}} f(z,k-1)\,dz \quad .$$

A function $f(x,k)$ on a domain $\mathcal{D}$ in $K^n \times \mathbb{Z}$ is sub-regular (resp., super-regular) if it is smooth, real-valued and the " $=$ " in (3.1) is replaced by " $\leqq$ " (respectively, " $\geqq$ ").

Remark. If $\mathcal{D} = K^n \times \mathbb{Z}$ the notion of regularity coincides with that of §1. Note also that $K^n \times \mathbb{Z}$ is unbounded but is connected and simple.

If $f(x,k)$ is a smooth function on a domain $\mathcal{D}$ its derived function $f'(x,k)$ may be determined as follows: If $(x,k) \in \mathcal{D} \approx \partial\mathcal{D}$ then $f'(x,k-1) = f(x,k-1) - f(x,k)$. If $(x,k-1) \in \mathcal{D}$ and $(x,k) \notin \mathcal{D}$ then $f'(x,k-1) = 0$. Our definition of regular, sub-regular and super-regular may be trivially reformulated:

Let f be a smooth function on a domain $\mathcal{D}$ and let $f'(x,k)$ be its derived function. Then

$$f \text{ is } \left\{\begin{matrix} \text{sub-regular} \\ \text{regular} \\ \text{super regular} \end{matrix}\right\} \text{ iff } \forall (x,k) \in \mathcal{D} \approx \partial\mathcal{D}, \int_{x+\mathfrak{B}^{-k}} f'(z,k-1)\,dz \left\{\begin{matrix} \geq 0 \\ = 0 \\ \leq 0 \end{matrix}\right. .$$

Note that if f is regular on a domain $\mathfrak{D}$ and $f'(z,k-1)$ is known for $z \in x + \mathfrak{P}^{-k}$, $(x,k) \in \mathfrak{D} \approx \partial\mathfrak{D}$ and if any one of the $1 + q^n$ values $\{f(x,k), f(x+a_s^k, k-1)\}$ (where $\{a_s^k\}$ are q^n coset representativ of $\mathfrak{P}^{-k+1}$ in $\mathfrak{P}^{-k}$) is also known then all the other values are determined. Thus,

Proposition (3.2). *If f is regular on a connected domain $\mathfrak{D}$ and f' is given, then f is determined up to an additive constant.*

Proposition (3.3). *If $f(x,k)$ is sub-regular on a bounded domain $\mathfrak{D}$, then* $\sup_{(x,k)\in\mathfrak{D}} f(x,k) = \sup_{(x,k)\in\partial\mathfrak{D}} f(x,k)$.

Proof. Since $\mathfrak{D}$ is bounded it will suffice to consider the special case where $\mathfrak{D} = (x + \mathfrak{P}^{-k}, k) \cup (x + \mathfrak{P}^{-k}, k-1)$, with $\partial\mathfrak{D} = (x + \mathfrak{P}^{-k}, k-1)$ The result now follows from the definition of sub-regularity.

Proposition (3.4). *If $\mathfrak{D}$ is a bounded domain and f and g are regular functions that agree on $\partial\mathfrak{D}$ then they agree on $\mathfrak{D}$.*

Proof. $f - g$ is regular so the supremum of $f - g$ is taken on $\partial\mathfrak{D}$ (take real and imaginary parts). Similarly for $g - f$ and so $f = g$ on $\mathfrak{D}$.

Proposition (3.5). (a) *If f is regular $|f|$ is sub-regular.* (b) *Linear combinations of regular functions are regular. If f and* g _are sub-regular_, $a,b \geq 0$ _then_ $af + bg$ _is sub-regular_, $f \vee g$ _is sub-regular_. (c) _If_ f _is sub-regular and_ Φ _is a non-decreasing convex function on an interval conaining the range of_ f _then_ $\Phi \circ f$ _is sub-regular_.

The proofs are left as an exercise.

Proposition (3.6). _If_ $f(x,k)$ _is sub-regular on_ $K^n \times \mathbb{Z}$ _then_ $R(\cdot,k) * f(\cdot,\ell) \geq f(\cdot,k \vee \ell)$.

Proof. Since f is smooth, if $k \leq \ell$, $R(\cdot,k) * f(\cdot,\ell) = f(x,\ell)$. The only case of interest is $k \geq \ell$, and that will follow if we check it for $\ell = k-1$. That is, $R(\cdot,k) * f(\cdot,k-1) \geq f(\cdot,k)$. But that is the definition of sub-regularity.

If a regular function $m(x,k)$ majorizes the function $f(x,k)$ on a domain $\mathcal{D}$ in $K^n \times \mathbb{Z}$ we say that m is a _regular majorant_ of f on $\mathcal{D}$. If (h, a regular majorant of f on $\mathcal{D}$) $\Longrightarrow$ $(h \geq m)$ then we say that m is a _least regular majorant_ of f on $\mathcal{D}$.

Convention. If $f \in \mathcal{S}'$ we identify f and the regular function $f(x,k)$. In view of (1.5) this is reasonable.

Theorem (3.7). _Suppose_ f _is a non-negative valued, sub-regular function on_ $K^n \times \mathbb{Z}$ _and suppose that_ $\sup_k \|f(\cdot,k)\|_p = A < \infty$ _for some_ p,

$1 \leq p \leq \infty$. Then $f(x,k)$ has a least regular majorant m on $K^n \times \mathbb{Z}$ and

(a) if $1 < p \leq \infty$, $m \in L^p$, $\|m\|_p = A$

(b) if $p = 1$, $m \in M$, $\|m\|_M = A$.

Proof. For $k,\ell \in \mathbb{Z}$ let $m_\ell(x,k) = (R(\circ,k) * f(\cdot,\ell))(x)$.

Since $f(x,\ell)$ is sub-regular, if $\ell \leq k$,

$m_\ell(x,k) = (R(\cdot,k) * f(\cdot,\ell))(x) \leq (R(\cdot,k) * f(\cdot,\ell-1))(x) = m_{\ell-1}(x,k)$.

That is, for $\ell \leq k$, $\{m_\ell(x,k)\}$ is a non-decreasing sequence as $\ell \longrightarrow -\infty$, and thus it has a limit.

Let $m(x,k) = \lim\limits_{\ell \to -\infty} m_\ell(x,k)$.

Moreover, for $\ell \leq k$ we see,

$f(x,k) \leq f(\cdot,\ell) * R(\circ,k) = m_\ell(x,k) \leq m(x,k)$, so m majorizes f. We now show that m is regular.

If ℓ is fixed we see that $m_\ell(x,k)$ is regular. By the Lebesgue monotone convergence theorem, $m(x,k) = \lim\limits_{\ell \to -\infty} m_\ell(x,k)$

$$= \lim_{\ell \to -\infty} q^{-kn} \int_{x+\mathfrak{P}^{-k}} m_\ell(z,k-1)\,dz = q^{-kn} \int_{x+\mathfrak{P}^{-k}} m(z,k-1)\,dz \ .$$

From (1.8) (a) there follows: $\|m_\ell(\cdot,k)\|_p \leq \|f(\circ,\ell)\|_p \leq A$. It is now clear (the monotone convergence theorem for $1 \leq p < \infty$, obviou for $p = \infty$) that $\|m(\cdot,k)\|_p \leq A$. But m majorizes f so $\sup\limits_k \|m(\cdot,k)\|_p =$

and by (1.9)(a) or (1.9)(b) we get $m \in L^p$ (resp, $m \in M$) and from (1.8) that $\|m\|_p = A$ (resp, $\|m\|_M = A$).

We now need to show that m is a least regular majorant. If $h(x,k)$ is a regular majorant of $f(x,k)$ then for $\ell \leq k$, $m_\ell(x,k) = (R(\cdot,k) * f(\cdot,\ell))(x) \leq (R(\cdot,k) * h(\cdot,\ell))(x) = h(x,k)$. Thus, $m(x,k) \leq h(x,k)$.

Let us identify K^n with $K^n \times \{-\infty\}$, the "boundary" of $K^n \times \mathbb{Z}$. For $\ell \geq 0$, $\ell \in \mathbb{Z}$, $z \in K^n$, let

$$\Gamma_\ell(z) = \{(x,k) \in K^n \times \mathbb{Z} : |x-z| \leq q^{k+\ell}\}$$

$$= \{(x,k) \in K^n \times \mathbb{Z} : x \in z + \mathfrak{B}^{-(k+\ell)}\}.$$

If $f(x,k)$ is defined on $K^n \times \mathbb{Z}$ we say that it has non-tangential limit L at $z \in K^n$ if for each non-negative integer ℓ, $\lim f(x,k) = L$ as (x,k) tends to $(z,-\infty)$ within $\Gamma_\ell(z)$. We write $\text{n.t.} \lim_{(x,k)\to z} f(z,k) = L$.

If f is smooth then the limit within $\Gamma_0(z)$ is the same as the "radial" limit, $\lim_{k \to -\infty} f(z,k)$.

Definition. If f is locally integrable we define the non-tangential Hardy-Littlewood maximal function for each non-negative integer ℓ,

$$M_\ell f(z) = \sup_{(x,k) \in \Gamma_\ell(z)} |f|(x,k).$$

Notice that $M_0 f = Mf$. It is also obvious that

(3.8) $\quad Mf(z) \leq M_\ell f(z), \quad \ell \geq 0, \quad \ell \in \mathbb{Z}$.

<u>Proposition (3.9)</u>. $\quad M_\ell f(z) \leq q^{n\ell} Mf(z)$.

<u>Proof</u>. Suppose $(x,k) \in \Gamma_\ell(z)$. Then,

$$|f|(x,k) = q^{-nk} \int_{x+\mathfrak{P}^{-k}} |f(y)|dy \leq q^{-nk} \int_{x+\mathfrak{P}^{-(k+\ell)}} |f(y)|dy$$

$$= q^{n\ell} q^{-n(k+\ell)} \int_{z+\mathfrak{P}^{-(k+\ell)}} |f(y)|dy = q^{n\ell}|f|(z,k+\ell).$$

Thus, $M_\ell f(z) = \sup_{(x,k)\in\Gamma_\ell(z)} |f|(x,k)$

$$\leq \sup_{k\in\mathbb{Z}} q^{n\ell}|f|(z,k+\ell)$$

$$= q^{n\ell} \sup_{k\in\mathbb{Z}} |f|(z,k) = q^{n\ell} Mf(z).$$

<u>Proposition (3.10)</u>. (a) <u>The operator</u> $f \rightarrow M_\ell f$ <u>is of type</u> (p,p), $1 < p \leq \infty$, <u>and of weak type</u> $(1,1)$ <u>for each non-negative integer</u> ℓ. <u>That is, there are constants</u> $A_{p\ell} > 0$, <u>independent of</u> f, <u>such that if</u> $f \in L^p$, $1 < p \leq \infty$, $\|M_\ell f\|_p \leq A_{p\ell}\|f\|_p$, <u>and for each</u> $s > 0$ <u>if</u> $f \in L^1$, $|\{x: M_\ell f(x) > s\}| \leq A_{1\ell}\|f\|_1 \, s^{-1}$.

(b) <u>If</u> f <u>is locally integrable, then</u> n.t. $\lim_{(x,k)\to z} f(x,k) = f(x)$.

Proof. For $\ell = 0$, (a) is (1.7) and (b) is III (1.13). For $\ell > 0$, (a) follows from the case $\ell = 0$ and (3.9). For (b) we see that the proof only depends on the fact that $f \to M_\ell f$ is of weak type (1,1) (See the proof of II (1.14)) and we get this from (a).

We construct a substitute for conformal mapping.

Definition. Let $\mathcal{D}$ be a simple domain in $K^n \times \mathbb{Z}$ and let $m = m(\mathcal{D})$. Suppose $f(x,k)$ is regular on $\mathcal{D}$, then the extension of f, $\bar{f}(x,k)$ to $K^n \times \mathbb{Z}$ is defined as follows:

$$\bar{f}(x,k) = \begin{cases} f(x,k), & (x,k) \in \mathcal{D} \\ f(x,\ell), & (x,\ell) \in \partial\mathcal{D} \text{ and } k \le \ell \\ \quad \text{If } (x,k) \notin \mathcal{D} \text{ or } (x,\ell) \notin \mathcal{D} \text{ for some } \ell \ge k \text{ then ,} & \\ 0 \quad , & k \le m \\ (R(\cdot,k) * \bar{f}(\cdot,m))(x), & k \ge m \end{cases}$$

Fact. If f is regular on the simple domain $\mathcal{D}$, then $\bar{f}$ is regular on $K^n \times \mathbb{Z}$.

4. Notes for Chapter IV

A general reference for §1 and §2 of this chapter is Taibleson [4] For §3 see Chao [1].

§1. Throughout this chapter, as well as in Chapters V and VI, we emphasize the analogy between regular functions on $K^n \times \mathbb{Z}$ and harmoni functions on $\mathbb{R}^n \times (0,\infty) = \mathbb{R}^{n+1,+}$, the euclidean upper half space. Wi that idea in hand we look at the definition of a regular function, (1.3 and see that it has two parts. In (i) we have a smoothness condition and in (ii) and (ii′) a substitute for the mean value property of harmonic functions.

The development then continues along the line worked out in Taibleson [6,Ch.II,pp.413-420], or in Stein [2] where the results are spread out through Chapters III, IV, V and VII.

The two parts of (1.3) can also be interpreted as showing that the sequence $\{f_k(\cdot)\}$, $f_k(x) = f(x,-k)$ is a martingale with respect to the sigma-algebras $\{\mathfrak{F}_k\}$, where $\mathfrak{F}_k$ is the sigma-algebra generated by the cosets of $\mathfrak{P}^k$. (1.3) (i) shows that f_k is measurable with respect to $\mathfrak{F}_k$, and (ii) shows that the expectation operator $E_k : f \rightarrow f(\cdot,-k)$ $(E_k f = E(f|\mathfrak{F}_k))$ has the martingale property: $E_k f_\ell = f_{\ell \wedge k}$. The deta of this last observation are worked out in (1.4), where we are maintaining the point of view that the result is the local field analogue of the semi-group property of the Poisson integral operator. For a general background see Doob [1] and for some developments more in the spirit of harmonic analysis see Burkholder [1] and [2] and Gundy [1].

From (1.8)(a) and (1.9)(b) we see that for a given p, $1 < p \leq \infty$, $f \in L^p$ iff there is an $M > 0$ such that $\|f(\cdot,k)\|_p \leq M$ for all $k \in \mathbb{Z}$. This result fails for p = 1 as we see from examining the regular function R(x,k). We note that $\|R(\cdot,k)\|_1 \equiv 1$ but R(x,k) is not the regularization of an L^1 - function.

§2. Theorem (2.2) of this section gives some local field variants of results in Taibleson [6,Ch.III] that are needed in the sequel.

Lemma (2.1) is the discrete analogue of Hardy's inequality (see Hardy, Littlewood and Polya [1,pp.239-246]), and the reader will note that the proof is totally trivial. If one takes the trouble to write the relations in Hardy's inequality as integrals on the multiplicative group: $\{(0,\infty),dt/t\}$, we see that Hardy's inequality has the same trivial proof.

§3. In this section we work out some more local field variants of the properties of harmonic functions in euclidean half-spaces. One of the more important of these is (3.7) which provides sufficient conditions for a sub-regular function to have a least regular majorant. For probability buffs we point out that sub-regular functions correspond to sub-martingales as well as being analogues of sub-harmonic functions

The definition of the extension of a function, regular on a simple domain to a regular function on $K^n \times \mathbb{Z}$, corresponds to the extension of functions on certain sub-domains of the disc to the entire disc by conformal mapping, as used in Zygmund [1,vol.II,Ch.XIV,§1]. It will be used in Chapter V §2 to study the boundary behaviour of regular functions. In the study of the "boundary behaviour" of martingales the corresponding extension is obtained by use of a stopping time.

Chapter V. The Littlewood-Paley function and some applications

In §1 we introduce the Littlewood-Paley functions $g_p(\cdot;f)$, and study the relation between the L^p properties of f and $g_s(\cdot;f)$. In §2 we introduce a truncated version of $g_2(\cdot;f)$, Sf, and study the local equivalence of the n.t. convergence of $f(x,k)$, n.t. boundedness of $f(x,k)$, and existence of $Sf(x)$. (n.t. = "non-tangential")

1. The Littlewood-Paley function

Definition. If $f(x,k)$ is regular on $K^n \times \mathbb{Z}$ we define the Littlewood-Paley functions $g_p(\cdot;f)$ by

$$g_\infty(x;f) = \sup_{k \in \mathbb{Z}} |f(x,k) - f(x,k-1)|$$

$$g_p(x;f) = \Big[\sum_{k \in \mathbb{Z}} |f(x,k) - f(x,k-1)|^p\Big]^{1/p}, \ 1 \leq p < \infty .$$

If F is the distribution to which $f(x,k)$ converges we will also write $g_p(x;F) = g_p(x;f)$.

Lemma (1.1). *If* $f \in L^2$ *then* $g(\cdot;f) \in L^2$ *and* $\|g_2(\cdot;f)\|_2 = \|f\|_2$.

Proof. Using IV (1.11) we get that

$$\int_{K^n} |f(x,k) - f(x,k-1)|^2 dx = \int |f(x,k)|^2 dx - \int f(x,k)\overline{f(x,k-1)}\,dx$$

$$- \int f(x,k-1)\overline{f(x,k)}\,dx + \int |f(x,k-1)|^2 dx$$

$$= \int |f(x,k-1)|^2 dx - \int |f(x,k)|^2 dx \ . \quad \text{Thus,}$$

$$\int \Big(\sum_t^T |f(x,k) - f(x,k-1)|^2\Big) dx = \sum_t^T \Big[\int |f(x,k-1)|^2 dx - \int |f(x,k)|^2 dx\Big]$$

$$= \int |f(x,t-1)|^2 dx - \int |f(x,T)|^2 dx \ .$$

From IV (1.7), $|f(x,T)|^2 \leq (Mf(x))^2 \in L^1$. By IV (1.10), $f(x,T) \to 0$ as $T \to \infty$. By the dominated convergence theorem, $\int |f(x,T)|^2 dx \to 0$ as $T \to \infty$. By IV (1.8), $\int |f(x,t-1)|^2 dx \to \|f\|_2^2$, as $t \to -\infty$.

From the monotone convergence theorem, $\Big[\sum_t^T |f(x,k) - f(x,k-1)|^2\Big]^{1/2}$ converges a.e. to an L^2 function, and L^2-norm equal to $\|f\|_2$. Thus, $g_2(x,f) = \Big[\sum_{k \in \mathbb{Z}} |f(x,k) - f(x,k-1)|^2\Big]^{1/2}$, exists a.e., is in L^2 and $\|g_2(\cdot;f)\|_2 = \|f\|_2$.

Lemma (1.2). *If* $f \in L^1(K^n)$ *then* $g_2(x;f)$ *exists for* a.e. x *and* $|\{x : g_2(x;f) > s\}| \leq A\|f\|_1 s^{-1}$ *for each* $s > 0$ *where* $A > 0$ *is independent of* f.

Proof. We use the decomposition of III (7.6) - (7.9) and the argument of III (7.12)(c). Thus, with $s > 0$ fixed we have $f = f_1^s + f_2^s$ with $f_2^s \in L^2$, $\|f_2^s\|_2^2 \leq s\, q^n \|f\|_1$. Further $f_1^s(x) = 0$

if $x \notin D_s$, where $|D_s| \le \|f\|_1\, s^{-1}$. Thus, $g_2(x;f_1^s) = 0$ if $x \notin D_s$ and $g_2(x;f_2^s) \in L^2$ with $\|g_2(\cdot;f_2^s)\|_2^2 \le q^n s\|f\|_1$.

We may argue as in III (7.12)(c) and get that $g_2(x;f)$ exists a.e. and further that,

$$|\{x : g_2(x;f) > s\}| \le |D_s| + |\{x: g_2(x;f_2^s) > s/2\}|$$

$$\le \|f\|_1\, s^{-1} + 4\, q^n\, s\|f\|_1\, s^{-2} = (1 + 4\, q^n)\|f\|_1\, s^{-1}.$$

<u>Lemma (1.3)</u>. <u>If</u> $1 < p \le 2$ <u>and</u> $f \in L^p$, <u>then</u> $g_2(\cdot;f) \in L^p$ <u>and there are constants</u> $C_p > 0$ <u>independent of</u> f, <u>such that</u>

$\|g_2(\cdot;f)\|_p \le C_p\|f\|_p$.

<u>Proof</u>. The map $f \to g_2(\cdot;f)$ is sub-linear. Use (1.1) and (1.2) and the Marcinkiewicz interpolation theorem.

<u>Remark</u>. The next result is the sticky part of the argument. There is a pretty argument using Rademacher functions (see §3), but we will sketch an argument along more classical lines.

<u>Lemma (1.4)</u>. <u>If</u> $p > 4$, $f \in L^p$, <u>then</u> $g_2(x;f) \in L^p$ <u>and there is a constant</u> $A_p > 0$ <u>independent of</u> f <u>such that</u> $\|g_2(\cdot;f)\|_p \le A_p\|f\|_p$.

<u>Proof</u>. We assume, for the sake of simplicity that f is real valued. We fix two integers, $-\infty < t < T < +\infty$.

We note that $\sum_t^T (f(x,k)-f(x,k-1))^2 \in L^{p/2}$. Let $h \in L^s$, $\|h\|_s = 1$, $(1/s) + (2/p) = 1$. Since $p > 4$, $1 < s < 2$, and and (1.3) applies to h. We wish to estimate the integral,

$$I_{tT} = \int_{K^n} \left\{ \sum_t^T (f(x,k)-f(x,k-1))^2 \right\} h(x)\,dx$$

$$= \sum_t^T \int_{K^n} (f(x,k)-f(x,k-1))^2 h(x)\,dx$$

$$= \sum_t^T \int_{K^n} (f(x,k)-f(x,k-1))^2 h(x,k-1)\,dx.$$

This last step is valid since if $g(x,k)$ is a smooth function and $h(x,k)$ is the regularization of h, then $\int g(x,k)h(x)\,dx = \int g(x,k)h(x,k)\,dx$, provided the integrals exist. A proof along the lines of IV (1.11)(a) is easily supplied.

If we split I_{tT} into two terms (writing $(f(x,k)-f(x,k-1))^2 = (f(x,k)-f(x,k-1))f(x,k)-(f(x,k)-f(x,k-1))f(x,k-1))$ sum each term by parts, and gather terms we obtain:

$$I_{tT} = -2\int \sum_{t+1}^{T-1} f(x,k)(f(x,k)-f(x,k-1))(h(x,k)-h(x,k-1))\,dx$$

boundary term

$$\left\{ \begin{aligned} &+ \int f^2(x,T)h(x,T-1)\,dx - 2\int f(x,T)f(x,T-1)h(x,T-1)\,dx \\ &+ 2\int f^2(x,t)h(x,t)\,dx - 2\int f(x,t-1)f(x,t)h(x,t-1)\,dx \\ &+ \int f^2(x,t-1)h(x,t-1)\,dx \quad . \end{aligned} \right.$$

Using IV (1.7) we see that the boundary term is dominated by,

$$8\int (Mf(x))^2 Mh(x)\,dx \le 8\|(Mf)^2\|_{p/2}\|Mh\|_s \le 8\ A_s\|Mf\|_p^2 \le 8\ A_s A_p^2\|f\|_p\ .$$

The main term is estimated using (1.3):

$$\left|\int\left\{\sum_{t+1}^{T-1} f(x,k)(f(x,k)-f(x,k-1))(h(x,k)-h(x,k-1))\right\}dx\right|$$

$$\le \int Mf(x)\Big(\sum_{t+1}^{T-1}(f(x,k)-f(x,k-1))^2\Big)^{\frac12}\Big(\sum_{t+1}^{T-1}|h(x,k)-h(x,k-1)|^2\Big)^{\frac12}dx$$

$$\le \|Mf\|_p\ \|g_2(\cdot;h)\|_s\ \Big\|\Big(\sum_{k=t+1}^{T-1}(f(\cdot,k)-f(\cdot,k-1))^2\Big)^{\frac12}\Big\|_p$$

$$\le\ A_p\ C_s\|f\|_p\ \Big\|\Big(\sum_t^T(f(\cdot,k)-f(\cdot,k-1))^2\Big)^{\frac12}\Big\|_p\ .$$

By the converse to Hölder's inequality,

$$\Big\|\Big(\sum_t^T(f(\cdot,k)-f(\cdot,k-1))^2\Big)^{\frac12}\Big\|_p^2$$

$$\le A'\Big\{\|f\|_p^2 + \|f\|_p\ \Big\|\Big(\sum_t^T(f(\cdot,k)-f(\cdot,k-1))^2\Big)^{\frac12}\Big\|_p\Big\}$$

where $A' > 0$ is given by $A' = \max[8A_p^2A_s,\ 2A_pA_s]$, and is independent of f, t and T. Hence, there is a $B_p > 0$, independent of f, t and T such that $\Big\|\Big(\sum_t^T(f(x,k)-f(x,k-1))^2\Big)^{\frac12}\Big\|_p \le B_p\|f\|_p$.

From the monotone convergence theorem we now get that $g_2(x;f)$ exists a.e., is in L^p and $\|g_2(\cdot;f)\|_p \le B_p\|f\|_p$, $B_p > 0$ independent of f.

Theorem (1.5). *If* $f \in L^p$, $1 \leq p < \infty$, *then* $g_2(x;f)$ *exists for a.e.* x. *If* $1 < p < \infty$ *then* $g_2(x;f) \in L^p$ *and there is a constant* $A_p > 0$ *independent of* f *such that* $\|g_2(\cdot;f)\|_p \leq A_p\|f\|_p$.

Proof. Existence for $p = 1$ is given by (1.2). Existence and boundedness for $1 < p \leq 2$ and $4 < p < \infty$ is given by (1.3) and (1.4) respectively. The result for $2 < p \leq 4$ now follows by interpolation.

Theorem (1.6). *If* $f \in L^p$, $2 \leq p \leq \infty$ *then* $g_p(\cdot;f) \in L^p$ *and there is a constant* $A_p > 0$ *independent of* f *such that* $\|g_p(\cdot;f)\|_p \leq A_p\|f\|_p$.

Proof. For $p = \infty$ we have, $g_\infty(x;f) = \sup_k|f(x,k) - f(x,k-1)|$ $\leq 2 \sup_k|f(x,k)| = 2\|f\|_\infty$.

Let A_p be the constants of $\|Mf\|_p \leq A_p\|f\|_p$, $1 < p \leq \infty$ and B_p the constants of $\|g_2(\cdot;f)\|_p \leq B_p\|f\|_p$, $1 < p < \infty$. Then for $2 \leq p < \infty$,

$$\|g_p(\cdot;f)\|_p = \left[\int \sum_{-\infty}^{+\infty} |f(x,k) - f(x,k-1)|^p dx\right]^{1/p}$$

$$= \left[\int \sum_{-\infty}^{+\infty} |f(x,k) - f(x,k-1)|^2 |f(x,k) - f(x,k-1)|^{p-2} dx\right]^{1/p}$$

$$\leq \left[\int 2^{p-2}(Mf)^{p-2}(g_2(x;f))^2 dx\right]^{1/p}$$

$$\leq 2^{(p-2)/p}\left[\|(Mf)^{p-2}\|_{p/(p-2)}\|g_2(\cdot;f))^2\|_{p/2}\right]^{1/p}$$

$$= 2^{(p-2)/p} \|Mf\|_p^{(p-2)/p} \|g_2(\cdot;f)\|_p^{2/p}$$

$$\leq 2^{(p-2)/p} A_p^{(p-2)/p} \|f\|_p^{(p-2)/p} B_p^{2/p} \|f\|_p^{2/p} = C_p \|f\|_p .$$

In the remainder of this section we will concern ourselves with converses to (1.5) and (1.6).

Lemma (1.7). *If* $f \in L^p(K^n)$, $h \in L^s(K^n)$, $(1/p) + (1/s) = 1$, $1 < p, s < \infty$, *then*

$$\int_{K^n} f(x)h(x)\,dx = \int_{K^n} \Big\{ \sum_{k\in\mathbb{Z}} (f(x,k)-f(x,k-1))(h(x,k)-h(x,k-1)) \Big\} dx.$$

Proof. Taking absolute values inside the summation sign on the right hand side, we get, $\int_{K^n} \{ \sum_{k\in\mathbb{Z}} | \qquad | \} dx$

$\leq \int_{K^n} g_2(x;f) g_2(x;h)\,dx \leq \|g_2(\cdot;f)\|_p \|g_2(\cdot;h)\|_s < \infty$. Thus, it will suffice to show that

$$\sum_{k\in\mathbb{Z}} \Big\{ \int_{K^n} (f(x,k)-f(x,k-1))(h(x,k)-h(x,k-1))\,dx \Big\} = \int_{K^n} f(x)h(x)\,dx,$$

as follows from Fubini's theorem.

As in (1.1)(using IV(1.7)) we have

$$\int_{K^n} (f(x,k) - f(x,k-1))(h(x,k) - h(x,k-1))\,dx$$

$$= \int_{K^n} f(x,k-1)h(x,k-1)\,dx - \int_{K^n} f(x,k)h(x,k) .$$

Thus, $\sum_t^T \int_{K^n} (f(x,k) - f(x,k-1))(h(x,k) - h(x,k-1))dx$

$$= \int_{K^n} f(x,t-1)h(x,t-1)dx - \int_{K^n} f(x,T)h(x,T)dx.$$

$|f(x,T)h(x,T)| \leq Mf(x)Mh(x) \in L^1$ (IV (1.7)), and $|f(x,T)h(x,T)| \to 0$ as $T \to \infty$ (IV (1.10)), so by the dominated convergence theorem $\int f(x,T)h(x,T) \to 0$ as $T \to \infty$. Using IV (1.8) and Hölder's inequality we get that $\int f(x,t-1)h(x,t-1)dx \to \int f(x)h(x)dx$. Thus

$$\sum_{k\in\mathbb{Z}} \left\{ \int_{K^n} \qquad dx \right\} = \lim_{\substack{t \to -\infty \\ T \to +\infty}} \sum_t^T \left\{ \int_{K^n} \qquad dx \right\}$$

$$= \lim_{\substack{t \to -\infty \\ T \to +\infty}} \left\{ \int_{K^n} f(x,t-1)h(x,t-1)dx - \int_{K^n} f(x,T)h(x,T)dx \right\}$$

$$= \int_{K^n} f(x)h(x)dx.$$

Theorem (1.8). Let $f(x,k)$ be a regular function on $K^n \times \mathbb{Z}$ such that (i) $f(x,k) \to 0$ as $k \to +\infty$ for each x and (ii) $g_2(\cdot;f) \in L^p$, $1 < p < \infty$. Then $f(x,k)$ is the regularization of a function $F \in L^p$ and there is a constant $A_p > 0$ independent of f such that $\|F\|_p \leq A_p \|g_2(\cdot;f)\|_p$.

__Proof__. Fix integers t,T such that $-\infty < t < T < +\infty$. Let $\bar{f} = f(\cdot,T) - f(\cdot,t)$. Then $\bar{f} \in L^p$ since,

$$\bar{f}(x) = \Big|\sum_{t+1}^{T} (f(x,k)-f(x,k-1))\Big| \le (T-t)^{\frac{1}{2}} g_2(x;f) \in L^p .$$

$$\bar{f}(x,k) = \begin{cases} f(x,k)-f(x,k) = 0 & , \ k \ge T \\ f(x,T)-f(x,k) & , \ t \le k \le T \\ f(x,T)-f(x,t) & , \ k \le t \end{cases} .$$

Thus, $\bar{f}(x,k)-\bar{f}(x,k-1) = \begin{cases} -(f(x,k) - f(x,k-1)), & t \le k \le T \\ 0 & , \text{otherwise}, \end{cases}$

and $g_2(x;\bar{f}) \le g_2(x;f)$.

Let $h \in L^s$, $(1/s) + (1/p) = 1$, $\|h\|_s = 1$. From (1.7),

$$\Big|\int_{K^n} \bar{f}(x)h(x)\,dx\Big| = \Big|\int_{K^n} \sum_{k\in\mathbb{Z}} (\bar{f}(x,k)-\bar{f}(x,k-1))(h(x,k)-h(x,k-1))dx\Big|$$

$$\le \int_{K^n} g_2(x;\bar{f})g_2(x;h)\,dx \le \|g_2(\cdot;\bar{f})\|_p \|g_2(\cdot;h)\|_s$$

$$\le A_s\|h\|_s\|g_2(\cdot;f)\|_p \le A_s\|g_2(\cdot;f)\|_p ,$$

where A_s is the constant of (1.5), and it is independent of f,t and T. Thus, $\|f(\cdot,T)-f(\cdot,t)\|_p = \|\bar{f}\|_p \le A_s\|g_2(\cdot;f)\|_p$, for all $t < T$.

Since $f(x,T) \to 0$ as $T \to \infty$, it follows from Fatou's lemma that $f(\cdot,t) \in L^p$ and $\|f(\cdot,t)\|_p \le A_s\|g_2(\cdot,f)\|_p$ for all $t \in \mathbb{Z}$.

From IV (1.9) f is the regularization of $F \in L^p$ and by IV (1.8), $\|F\|_p = \sup_t \|f(\cdot,t)\|_p \leq A_s \|g_2(\cdot;f)\|_p$.

<u>Theorem (1.9)</u>. <u>Suppose</u> $f(x,k)$ <u>is a regular function on</u> $K^n \times \mathbb{Z}$ <u>such that</u> (i) $f(x,k) \to 0$ <u>as</u> $k \to +\infty$ <u>for each</u> x <u>and</u> (ii) $g_p(\cdot;f) \in L^p$, $1 \leq p \leq 2$. <u>Then</u> $f(x,k)$ <u>is the regularization of a function</u> $F \in L^p$ <u>and there is a constant</u> $A_p > 0$ <u>independent of</u> f <u>such that</u> $\|F\|_p \leq A_p \|g_p(\cdot;f)\|_p$.

<u>Proof</u>. If $1 < p \leq 2$ the result follows almost exactly as in the proof of (1.8). We only need to notice that $|f(x,T)-f(x,t)| \leq (T-t)^{1/p'} g_p(x;f) \in L^p$ and using (1.7) obtain

$$\left|\int \bar{f}(x)h(x)dx\right| \leq \left|\int g_p(x;\bar{f})g_{p'}(x;h)dx\right| \leq \|g_p(\cdot;f)\|_p \|g_{p'}(\cdot;h)\|_{p'} ,$$

$\bar{f} = f(\cdot,T)-f(\cdot,t)$, $h \in L^{p'}$. One uses (1.6) and the proof goes through.

This proof breaks down for $p = 1$. For that case we compute directly. We have that $f(x,k) \to 0$ as $k \to \infty$ and $\sum_{k\in\mathbb{Z}} |f(x,k)-f(x,k-1)| = g_1(x;f) \in L^1$. Thus,

$$\|f(x,T)-f(x,t)\|_1 = \left\|\sum_{t+1}^{T} (f(x,k)-f(x,k-1))\right\|_1$$

$$\leq \int\left\{\sum_{t+1}^{T} |f(x,k)-f(x,k-1)|\right\}dx \leq \|g_1(\cdot;f)\|_1 .$$

Since $f(x,T) \to 0$ as $T \to \infty$, Fatou's lemma implies that $\|f(x,t)\|_1 \le \|g_1(\cdot;f)\|_1$ for all $t \in \mathbb{Z}$.

From Fubini we have $\sum_{k\in\mathbb{Z}} \|f(\cdot,k)-f(\cdot,k-1)\| = \|g_1(\cdot,f)\|_1$.

Thus, if $t_1 < t_2$, $\|f(\cdot,t_2)-f(\cdot,t_1)\|_1 \le \sum_{t_1+1}^{t_2} \|f(\cdot,k)-f(\cdot,k-1)\|_1 \to 0$

as $t_1,t_2 \to -\infty$. Hence $\{f(\cdot,k)\}$ is bounded in L^1 and is Cauchy in L^1 as $k \to -\infty$. From IV (1.9) we see that f is the regularization of a function $F \in L^1$, and from IV (1.8) $\|F\|_1 = \sup_k \|f(\cdot,k)\|_1$ $\le \|g_1(\cdot;f)\|_1$.

2. Local equivalence of non-tangential convergence, non-tangential boundedness and the existence of Sf

Definition. If $f(x,k)$ is regular on $K^n \times \mathbb{Z}$ then

$$Sf(x) = \left(\sum_{-\infty}^{0} |f(x,k)-f(x,k-1)|^2\right)^{1/2}.$$

Note that $Sf(x) \le g_2(x;f)$ and is a truncated version of the Littlewood-Paley function.

Definition. Let F be a measurable subset of K^n. Then $z \in F$ is a point of density of F if

$$\lim_{k\to\infty} |(z+\mathfrak{P}^k) \cap F| \big/ |z+\mathfrak{P}^k| = 1 .$$

Proposition (2.1). Almost every point of a measurable subset F is a point of density of F.

Proof. The ratio is simply the regularization of the characteristic function of F. Now apply III (1.14).

Lemma (2.2). *Let ℓ be a positive integer, and suppose $\{(x_j,k_j)\}_{j=1}^{\infty} \subset \Gamma_{\ell}(z)$, with $(x_j,k_j) \rightarrow (z,-\infty)$ as $j \rightarrow \infty$. If z is a point of density of F, then there is a $J(z) \geq 0$ such that $j \geq J(z)$ implies $(x_j,k_j) \in \cup_{y\in F}\Gamma_0(y)$.*

Proof. Let $z \in F$, a point of density of F be fixed. If the result fails then there is a sequence $\{(x_j,k_j)\}$ such that $\{k_j\} \rightarrow -\infty$, $\{x_j\} \rightarrow z$, $(x_j,k_j) \in \Gamma_{\ell}(z)$ for all j and $(x_j,k_j) \notin \cup_{y\in F}\Gamma_0(y)$ for all j.

$$(x_j,k_j) \notin \cup_{y\in F}\Gamma_0(y) \quad \text{iff} \quad y \notin x_j + \mathfrak{P}^{-k_j} \quad \text{all } y \in F \text{ and } j .$$

This implies $(x_j + \mathfrak{P}^{-k_j}) \cap F = \emptyset$ all j.

Since $(x_j,k_j) \in \Gamma_{\ell}(z)$, $x_j \in z + \mathfrak{P}^{-(k_j+\ell)}$, so $x_j + \mathfrak{P}^{-k_j} \subset z + \mathfrak{P}^{-(k_j+\ell)}$

Let $E_j = (z + \mathfrak{P}^{-(k_j+\ell)}) \sim (x_j + \mathfrak{P}^{-k_j})$. Then $|E_j| = q^{n(k_j+\ell)} - q^{nk_j} = q^{nk_j}(q^{n\ell}-1)$.

Let ξ_F be the characteristic function of F. Then

$$\xi_F(z,k_j+\ell) = q^{-n(k_j+\ell)} \int_{y\in z+\mathfrak{P}^{-(k_j+\ell)}} \xi_F(y)\,dy$$

$$= q^{-n(k_j+\ell)}|F \cap (z+\mathfrak{P}^{-(k_j+\ell)})| = q^{-n(k_j+\ell)}|F \cap E_j|$$

$$\leq q^{-n(k_j+\ell)}|E_j| = q^{-n\ell}(q^{n\ell}-1) = 1 - q^{-n\ell} < 1 ,$$

which is a contradiction, since $\xi_F(z,k_j+\ell) \longrightarrow 1$ as $j \longrightarrow \infty$.

Notation. If A and B are measurable subsets of K^n we say that A is equivalent to B if $|A \Delta B| = 0$ and that $A \underset{\sim}{\subseteq} B$ if if $|A \sim B| = 0$. Clearly, $A \underset{\sim}{\subseteq} B$, $B \underset{\sim}{\subseteq} A$ iff A is equivalent to B and $A \subset B$ implies that $A \underset{\sim}{\subseteq} B$.

Theorem (2.3). *If $f(x,k)$ is regular on $K^n \times \mathbb{Z}$ then the following four sets are equivalent.*

$A = \{x \in K^n : \lim_{k \to -\infty} f(x,k) \text{ exists}\}$

$B = \{x \in K^n : \text{n.t. } \lim_{(z,k) \to x} f(z,k) \text{ exists}\}$

$C = \{x \in K^n : \sup_{k \leq 0} |f(x,k)| < \infty\}$

$D = \{x \in K^n : Sf(x) < \infty\}$.

Proof. $B \subset A \subset C$ is trivial. The outline of the proof is to show $C \underset{\sim}{\subseteq} B$, $C \underset{\sim}{\subseteq} D$ and then $D \underset{\sim}{\subseteq} B$.

Let C be the set where $f(x,k)$ is radially bounded. We want to show that f has non-tangential limits a.e. in C and that $Sf(x)$ exists a.e. in C. We may assume that C is contained in a coset of $\mathfrak{D}$, and for notational convenience we will assume $C \subset \mathfrak{D}$. Now let $E_M = \{x \in \mathfrak{D} : \sup_{k \leq 0} |f(x,k)| \leq M\}$. Since $C = \lim_{M \to \infty} E_M$ we may assume that $C = E_M$ for M a positive integer.

For $m \leq 0$, let $\Gamma_1^m(x) = \{(z,k) \in \Gamma_1(x) : k \leq m\}$. From (2.2) it follows that for a.e. $x \in E_M$ there is an integer $\ell(x)$ such that $\Gamma_1^{\ell(x)}(x) \subset \cup_{y \in E_M} \Gamma_0(y)$.

Let $E_M^m = \{x \in E_M : \Gamma_1^m(x) \subset \cup_{y \in E_M} \Gamma_0(y)\}$.

Since $\{E_M^m\} \uparrow$ as $m \to -\infty$ and the fact that $\cup_{m \leq 0} E_M^m$ is equivalent to E_M we may now assume that $C = E_M^m$.

<u>Claim:</u> $\mathfrak{A} = \cup_{x \in E_M^m} \Gamma_1^m(x)$ <u>is a simple domain</u> in $K^n \times \mathbb{Z}$ and $\partial\mathfrak{A} = \mathfrak{A} \approx \cup_{x \in E_M^m} \Gamma_0^m(x)$.

<u>Proof of Claim</u>. Refer to the definition of domains in IV, §3. ? $(z,k) \in \mathfrak{A} \Longrightarrow (z + \mathfrak{P}^{-k},k) \subset \mathfrak{A}$? : $(z,k) \in \mathfrak{A}$ implies $(z,k) \in \Gamma_1(x)$, or $(z,k) \in (x + \mathfrak{P}^{-(k+1)},k) \subset \mathfrak{A}$, but $x + \mathfrak{P}^{-(k+1)} = z + \mathfrak{P}^{-(k+1)}$. ? $(z,k) \in \mathfrak{A}$ and $(z,k-1) \in \mathfrak{A} \Longrightarrow (z+\mathfrak{P}^{-k},k-1) \subset \mathfrak{A}$? From the first part of the proof $(z,k-1) \in \mathfrak{A} \Longrightarrow (z+\mathfrak{P}^{-k},k-1) \subset \mathfrak{A}$.

We now check that $\partial\mathcal{D} = \mathcal{S} \sim \bigcup_{x\in E_M^m} \Gamma_0^m(x)$. We want to show that if $(z,k) \in \Gamma_1^m(x)$ for some $x \in E_M^m$ then $(z,k) \notin \Gamma_0(y)$ for any $y \in E_M^m$ iff $(z,k-1) \notin \Gamma_1^m(y)$ for any $y \in E_M^m$. But $(z,k-1) \in \Gamma_1^m(y)$ iff $(z,k) \in \Gamma_0^m(y)$ provided $k - 1 \neq m$. But $(z,k) \in \Gamma_1^m(x)$ for some x, so $k \leq m$ and so $k-1 < m$. Thus $\partial\mathcal{D}$ is as advertised.

? $(z,k) \in \partial\mathcal{D} \Longrightarrow (z,\ell) \notin \mathcal{D}$ for all $\ell < k$. ? : From our construction of $\mathcal{D}$ if $(z,\ell) \in \mathcal{D}$ then $(z,t) \in \mathcal{D}$, $\ell \leq t \leq m$. Since $(z,k) \in \partial\mathcal{D}$, we have that $k \leq m$, $(z,k-1) \notin \mathcal{D}$. If for some ℓ, $\ell < k-1 < m$ we had $(z,\ell) \in \mathcal{D}$ we would also have $(z,k-1) \in \mathcal{D}$, a contradiction. Thus $\mathcal{D}$ is a domain.

Now notice that if for some k, $(x,k) \in \mathcal{D}$ then $m(x) = m$, otherwise $m(x) = \infty$ so $m(\mathcal{D}) = m$ and $\mathcal{D}$ is simple.

This proves the claim.

We now have that $\mathcal{D} = \bigcup_{x\in E_M^m} \Gamma_1^m(y)$ is a simple domain, f is bounded by M on $\mathcal{D}$. Let $\overline{f}(x,k)$ be the extension of f to $K^n \times \mathbb{Z}$ defined at the end of IV §3. From that construction it is immediate that $|\overline{f}(x,k)| \leq M$ for all $(x,k) \in K^n \times \mathbb{Z}$. Thus, $\overline{f}(x,k)$ is the regularization of a function $F \in L^\infty$ (IV (1.9)). From IV (3.10) we have that, n.t. $\lim_{(z,k)\to x} f(z,k)$ exists for a.e. $x \in K^n$ and hence for a.e. $x \in E_M^m$. Let us assume that x is a point of density of E_M^m. Suppose $(z_s,k_s) \rightarrow (x,-\infty)$ in $\Gamma_\ell(x)$. By (2.2) the sequence is eventually in $\cup_{y\in E_M^m} \Gamma_0^m(y) \subset \mathcal{D}$. Hence for s large enough, $\bar{f}(z_s,k_s) = f(z_s,k_s)$ so $\lim_s f(z_s,k_s)$ exists for a.e. $x \in E_M^m$ if (z_s,k_s) approaches x in $\Gamma_\ell(x)$ and thus, n.t. $\lim_{(z,k)\to x} f(z,k)$ exists a.e. in E_M^m. This shows that $C \underset{\sim}{\subseteq} B$.

From our construction of $\bar{f}(x,k)$ we see that F is supported compactly. Since F is bounded, $F \in L^2$ and so $S\bar{f}(x) \leq g_2(x;\bar{f})$ exists a.e. on K^n, and in particular a.e. on E_M^m. If $x \in E_M^m$, then $\bar{f}(x,k)$ and $f(x,k)$ can only differ if $m < k \leq 0$, so $S\bar{f}(x)$ and $Sf(x)$ differ by a finite number of terms and $Sf(x) < \infty$ iff $S\bar{f}(x) < \infty$ and this shows $C \underset{\sim}{\subseteq} D$.

We now proceed to showing $D \underset{\sim}{\subseteq} B$. The essential idea is to use IV (1.8) for $p = 2$.

As before we construct $E_M^m = \{x \in E_M : \Gamma_1^m(x) \subset \cup_{y\in E_M} \Gamma_0(y)\}$. This time $E_M = \{x \in \mathcal{O} : Sf(x) \leq M\}$. We will show that $f(x,k)$ has n.t. limits a.e. on E_M^m.

It follows as before that $\mathcal{D} = \cup_{x\in E_M^m} \Gamma_1^m(x)$ is a simple domain. The restriction of $f(x,k)$ to $\mathcal{D}$ is regular and we let $\bar{f}(x,k)$ be the extension of $f(x,k)$ to $K^n \times \mathbb{Z}$.

Let us assume that $m < 0$. Then $\bar{f}(x,k) = 0$ if $x \notin \mathfrak{D}$, and $Sf(x) = 0$ if $x \notin \mathfrak{D}$. We now consider three cases for $x \in \mathfrak{D}$:

$x \in E_M^m$: Then $(x,k) \in \mathcal{D}$ for $-\infty < k \leq m$, so $\bar{f}(x,k) = f(x,k)$ if $-\infty < k \leq m$ and $\left[\sum_{k=-\infty}^{m} |\bar{f}(x,k)-\bar{f}(x,k-1)|^2\right]^{\frac{1}{2}} \leq Sf(x) \leq M$.

$(x,k) \notin \mathcal{D}$ for any k: Then $\bar{f}(x,k) = 0$, $-\infty < k \leq m$, and $\left[\sum_{k=-\infty}^{m} |\bar{f}(x,k)-f(x,k-1)|^2\right]^{\frac{1}{2}} = 0$.

$x \notin E_M^m$, but $(x,\ell) \in \partial\mathcal{D}$ for some ℓ: Thus $\bar{f}(x,k) = f(x,k)$ if $\ell \leq k \leq m$, but, $\bar{f}(x,k) = f(x,\ell)$ if $-\infty < k \leq \ell$. Thus,

$$\left[\sum_{k=-\infty}^{m} |\bar{f}(x,k)-\bar{f}(x,k-1)|^2\right]^{\frac{1}{2}} = \left[\sum_{k=\ell+1}^{m} |f(x,k)-f(x,k-1)|^2\right]^{\frac{1}{2}} .$$

Since $(x,\ell) \in \partial\mathcal{D}$, $(x,\ell) \in \Gamma_1(z)$ for some $z \in E_M^m$. Thus, $(x,\ell) \in \Gamma_0(y)$ for some $y \in E_M$. For $k \geq \ell$, $f(x,\ell) = f(y,\ell)$. Thus,

$$\left[\sum_{k=-\infty}^{m} |\bar{f}(x,k)-\bar{f}(x,k-1)|^2\right]^{\frac{1}{2}} \leq \left[\sum_{k=\ell+1}^{m} |f(y,k)-f(y,k-1)|^2\right]^{\frac{1}{2}}$$

$$\leq Sf(y) \leq M .$$

Thus, for $x \in \mathfrak{D}$, $\left[\sum_{k=-\infty}^{m} |\bar{f}(x,k)-\bar{f}(x,k-1)|^2\right]^{\frac{1}{2}} \leq M$. The difference between this expression on $S\bar{f}(x)$ is bounded by $\left[\sum_{m+1}^{0} |\bar{f}(x,k)-\bar{f}(x,k-1)|^2\right]^{\frac{1}{2}}$ which is bounded by a constant that depends only on m and $\sup_{x\in\mathfrak{D}}|\bar{f}(x,m)| = A$, so that $S\bar{f}(x)$ is a bounded function with support on $\mathfrak{D}$. Now notice that $\bar{f}(x,0) = B\,R(x,0)$, where $|B| \leq A$. Thus, $\bar{f}(x,k) = B\,R(x,k)$, $k \geq 0$, and $\bar{f}(x,k) \rightarrow 0$ as $k \rightarrow +\infty$, $S\bar{f} \in L^2(K^n)$ and

$$g_2(x;\bar{f}) \leq S\bar{f}(x) + A\left[\sum_{k=1}^{\infty} |R(x,k)-R(x,k-1)|^2\right]^{\frac{1}{2}} .$$

Arguing as in (1.1) we see that the L^2 norm of this last term is A. Hence $g_2(x;\bar{f}) \in L^2(K^n)$. From IV (1.8) we see that $\bar{f}(x,k)$ is the regularization of a function in L^2 so the non-tangential limits of $\bar{f}$ exist a.e. We then argue as in the proof of $C \underset{\sim}{\subseteq} B$ and get that $f(x,k)$ has non-tangential limits a.e. on E_M^m. Thus, $D \underset{\sim}{\subseteq} B$ and the proof is complete.

<u>Remark</u>. It is trivial to see that the set where f is non-tangentially bounded is equivalent to sets defined in (2.3). Let E be that set. Then $B \subset E \subset C$. We could also use IV (3.9).

3. Notes for Chapter V

A general reference for §1 is Taibleson [4, §3]; for §2 see Chao [1].

§1. The term "Littlewood-Paley (L-P) function" refers to a rather large collection of different operator defined functions. The first of these was defined by Paley [1] as follows: Let f be a function on the Walsh-Paley group (2^{ω}), and let $S_n f(x)$ be the nth partial sum of its Fourier series. Then let

$$(3.1) \qquad g(x;f) = \left[\sum_{k=0}^{\infty} |S_{2^{k+1}} f(x) - S_{2^k} f(x)|^2\right]^{\frac{1}{2}} .$$

At about the same time (1931-32), Littlewood and Paley [1] were introducing another function defined as follows: Let f be a function on the circle, and let $f(r,\theta)$ be its Poisson integral, $0 \leq r < 1$, $0 \leq \theta < 2\pi$. Then let

$$(3.2) \qquad g(\theta;f) = \left[\int_0^1 (1-r)|f_r(r,\theta)|^2 dr\right]^{\frac{1}{2}} .$$

This L-P function was designed to carry much the same information as the L-P function in (3.1), if the partial sums in that definition were those of ordinary trigonometric Fourier series. A discussion of these heuristic and formal connections can be found in Zygmund [1, Ch XV, §1]. The discussion on pp. 222-223 is particularly illuminating.

In 1958 Stein [1] gave the following definition for functions, f, defined on R^n. Let f(x,y) be the Poisson integral of f on $R^n \times (0,\infty)$. Then let

$$g(x;f) = \left[\int_0^\infty y|f_y(x,y)|^2 dy\right]^{\frac{1}{2}} \quad . \tag{3.3}$$

It should be noted that with g defined as in (3.1), (3.2) or (3.3) we get conclusions as in (1.6) and (1.8).

Extending these ideas further, Hörmander [1] defined a class of L-P functions. One of these can be found in Taibleson [6,pp.457-460] where we have a heat equation version of the L-P function on R^n.

As we observed earlier (see IV §4 and II (6.7)) the family of operators $f \rightarrow S_{2^n}f$, for f on 2^ω are expectation operators and $\left\{S_{2^n}f\right\}_{n=0}^{\infty}$ is a martingale. Thus, it is not surprising that we also have martingale versions of the L-P theory. For a treatment of this subject see Gundy [2] and Burkholder [2]. See also Stein [3] where he gives a development of L-P theory on compact Lie groups and on martingales, as well as a "general L-P theory" for suitably nice semi-groups of operators.

It is a curious fact that the L-P function we describe in §1 was designed to be an "obvious" extension of the L-P function in Stein [1], in the sense that we replace $yf_y(x,y)$ with $f(x,k)-f(x,k-1)$ and Haar measure dy/y on $(0,\infty)$ with Haar measure on the integers as an additive group ($(0,\infty)$ and $\mathbb{Z}$ are images of the corresponding fields under the valuation operator). However, if we view the situation probabilistically, it turns out to agree with the L-P function defined on martingales. When we restrict the L-P function defined in §1 to the ring of integers in the 2-series field (which is the Walsh-Paley group, $2^{(\omega)}$), it turns out to be essentially identical with the function defined by Paley in 1932.

§2. Consider the L-P function defined in (3.2). In that integral defining g, if we replace the integral along the "radial ray", $0 \leq r < 1$, with an integral (suitably normalized) along a "conical region" projecting into the unit disc (see Zygmund [1, pp. 207-208]), we get the function $s(\theta;f)$, known as the Lusin function or area integral. It is well known that $s(\theta)$ dominates $g(\theta)$ pointwise (Zygmund [1,p.210]). Much more is true in our setting. In fact, the two functions agree since if $f(x,k)$ is regular, it is constant on sets of the form $(x + \mathfrak{P}^{-k},k)$ and so a "suitably normalized" integral on a "conical region" extending into the interior (namely, $\Gamma_0(x)$) agrees with the radial version of g.

Thus the well known result for the disc that, up to sets of measure zero, if f is a harmonic function in the disc, the set of points where f has non-tangential limits, where f is non-tangentially bounded, or where $s(\theta;f) < \infty$, are the same set, and yields (2.3) as its obvious generalization. The original result (in its various parts) is due to Marcinkiewicz and Zygmund, Spencer, and Plessner. Details and specific references can be found in the notes to Chapter XIV in Zygmund [1]. The R^n version of these results is due to Calderón [1] and [2]. See also Stein's paper (Acta Math. 106 (1961 137-174) where it is shown that the set where the area integral is finite is equivalent to the set where the non-tangential limit is finite.

Chapter VI. Multipliers and singular integral operators

In §1 and §2 we give a brief overview of the theory of L^p-multipliers on K^n and in §3 give an important application of the multiplier theory for Fourier series. In §4 we record, without proofs, the important facts about the singular integral theory.

1. Multipliers

We say that the measurable function m is a multiplier on L^p $(1 \leq p \leq \infty)$ if $(m\hat{\varphi})^{\vee} \in L^p$ for all $\varphi \in \mathcal{S}$ and there is a constant $A > 0$, independent of φ such that
$\|(m\hat{\varphi})^{\vee}\|_p \leq A\|\varphi\|_p$.

Since the map $\varphi \rightarrow (m\hat{\varphi})^{\vee}$ commutes with translations it follows (III (7.5)) that m is also a multiplier on $L^{p'}$ $((1/p) + (1/p') = 1)$, and so by the Riesz-Thorin interpolation theorem m is a multiplier on L^2. From III (10.6) we see that $m \in L^{\infty}$. Even if we had started with $m \in \mathcal{S}'$, by now we would have $m \in L^{\infty}$, so we assume that property as part of the formal definition of a multiplier.

Definition. m is a multiplier on L^p, $1 \leq p \leq \infty$ if $m \in L^{\infty}$ and there is a constant $A > 0$ independent of $\varphi \in \mathcal{S}$ such that $(m\hat{\varphi})^{\vee} \in L^p$ for all $\varphi \in \mathcal{S}$ and $\|(m\hat{\varphi})^{\vee}\|_p \leq A\|\varphi\|_p$.

We note here that multipliers are the Fourier transforms of the elements of T^p_p described in III §10.

Let $m_k = m\Phi_k$ (so $(m_k)^\vee = m^\vee(\cdot,k)$). Since $\varphi \in \mathcal{S}$ has compact support we see that $(m_k\hat{\varphi})^\vee = (m\,\hat{\varphi})^\vee$ for k small enough. Hence if we wish to show that m is a multiplier on L^p we only need to show that $m \in L^\infty$ and that there is a constant $A > 0$, independent of $k \in \mathbb{Z}$ and $\varphi \in \mathcal{S}$, such that $\|(m_k\hat{\varphi})^\vee\|_p \leq A\|\varphi\|_p$.

Observe now that since $m \in L^\infty$, then $m_k \in L^\infty \cap L^1 \subset L^2$ so $m_k{}^\vee \in L^2 \cap L^\infty$. Thus, $(m_k\hat{\varphi})^\vee = (m_k)^\vee * \varphi$, where the convolution is defined as an integral in the ordinary manner.

<u>Theorem (1.1)</u>. <u>Suppose</u> $m \in L^\infty$ <u>and there are</u> $B > 0$, $\epsilon > 0$, <u>independent of</u> $\ell \in \mathbb{Z}$ <u>such that</u>

$$(1.2) \qquad \int_{|y|<q^\ell} \int_{|x|=q^\ell} |m(x+y)-m(x)|^2 dx \, \frac{dy}{|y|^{2n+\epsilon}} \leq B^2 q^{-\epsilon\ell} .$$

<u>Then</u> m <u>is a multiplier on</u> L^p, $1 < p < \infty$. <u>In particular there are constants</u> $A > 0$, $C_p = C_{p'} > 0$ $(1/p + 1/p' = 1)$ <u>with</u> $C_p = O(p)$ $p \to \infty$, <u>where</u> A <u>depends only on</u> n, ϵ, B <u>and</u> $\|m\|_\infty$, C_p <u>depends only on</u> p <u>such that</u> $(m\hat{\varphi})^\vee \in L^p$, $\|(m\hat{\varphi})^\vee\|_p \leq AC_p\|\varphi\|_p$, $\forall\, \varphi \in \mathcal{S}$.

Example. If m is a bounded radial function the integral in (1.2) is identically zero, so such functions are trivially seen to be multipliers. Instances of interest are $\hat{G}^{\alpha}(\xi) = (\max[1,|\xi|])^{-\alpha}$, and $|\xi|^{-\alpha}$, for $\mathrm{Re}(\alpha) = 0$.

Claim. If we can show that there is an $A > 0$ (as in (1.1)) independent of $k \in \mathbb{Z}$ such that for all $\varphi \in \mathcal{S}$ and $s > 0$,

$$(1.3) \qquad |\{x : |(m_k)^{\vee} * \varphi(x)| > s\}| \leq A\|\varphi\|_1 \, s^{-1}$$

then (1.1) will follow.

To see this note that to establish (1.1) we need only show that there are constants A and C_p as in (1.1) that are independent of $k \in \mathbb{Z}$ such that $\|(m_k)^{\vee} * \varphi\|_p \leq A\, C_p\|\varphi\|_p$ for all $\varphi \in \mathcal{S}$.

But $\|(m_k)^{\vee} * \varphi\|_2 \leq \|m\|_{\infty}\|\varphi\|_2$, so we have the result for $p = 2$. But if (1.3) holds then the map is of weak type (1,1) with constant A. From the Marcinkiewicz interpolation theorem the maps are bounded in L^p with norms $A\,C_p$, $C_p = O(1/(p-1))$ as $p \to 1$. An application of III (7.5) completes the proof of the claim.

We will now work towards a proof of (1.3). This is stated as Lemma (1.9).

Lemma (1.4). If $\varphi \in \mathscr{S}$, $m \in L^{\infty}$, $\|(m_k)^{\vee} * \varphi\|_2 \leq \|m\|_{\infty} \|\varphi\|_2$.

Stated here for convenience.

Lemma (1.5). If $g \in L^2$, α real, then

$$q^{-\alpha} \int_{K^n} |\xi|^{\alpha} |\hat{g}(\xi)|^2 d\xi = \sum_{k \in \mathbb{Z}} q^{-k\alpha} \|g(\cdot,k) - g(\cdot,k-1)\|_2^2$$

in the sense that if either is finite so is the other and they are equal.

Proof. $(g(\cdot,k) - g(\cdot,k-1))^{\wedge} = \hat{g}(\Phi_k - \Phi_{k-1})$. Thus,

$$\|g(\cdot,k) - g(\cdot,k-1)\|_2^2 = \int_{|\xi| = q^{-k+1}} |\hat{g}(\xi)|^2 d\xi \text{ , and}$$

$$q^{-k\alpha} \|g(\cdot,k) - g(\cdot,k-1)\|_2^2 = q^{\alpha} \int_{|\xi| = q^{-k+1}} |\xi|^{\alpha} |\hat{g}(\xi)|^2 d\xi \text{ ,}$$

and the lemma follows.

Lemma (1.6). If $g \in L^2$, $\alpha > 0$, there is a constant $A_{\alpha} > 0$, independent of g such that

$$\sum q^{-k\alpha} \|g(\cdot,k) - g(\cdot,k-1)\|_2^2$$

$$\leq A_{\alpha} \int_{y \in K^n} \int_{x \in K^n} |g(x+y) - g(x)|^2 dx \frac{dy}{|y|^{n+\alpha}} \text{ .}$$

Proof. This follows from the equivalence of the norms "A" and "B" in IV (2.2).

Lemma (1.7). *If* $g \in L^2(K^n)$, $\epsilon > 0$ *and*

$$\iint |g(x+y) - g(x)| \, dx \, |y|^{-(2n+\epsilon)} dy = B^2 ,$$

then for every $k \in \mathbb{Z}$, $\displaystyle\int_{|\xi| \geq q^k} |\hat{g}(\xi)| d\xi \leq A_{n\epsilon} B \, q^{-k\epsilon/2}$,

where $A_{n\epsilon} > 0$ *depends only on* n *and* ϵ.

Proof.
$$\int_{|\xi| \geq q^k} |\hat{g}(\xi)| d\xi = \int_{|\xi| \geq q^k} (|\xi|^{(n+\epsilon)/2} \hat{g}(\xi)) |\xi|^{-(n+\epsilon)/2} d\xi$$

$$\leq \left[\int_{K^n} |\xi|^{n+\epsilon} |\hat{g}(\xi)|^2 d\xi \right]^{\frac{1}{2}} \left[\int_{|\xi| \geq q^k} |\xi|^{-(n+\epsilon)} d\xi \right]^{\frac{1}{2}}$$

$$\leq A_{n\epsilon} \, q^{-k\epsilon/2} \left[\sum_{\ell} q^{-\ell(n+2)} \|g(\cdot,\ell) - g(\cdot,\ell-1)\|_2^2 \right]^{\frac{1}{2}}$$

$$\leq A_{n\epsilon} \, q^{-k\epsilon/2} \left[\iint |g(x+y) - g(x)|^2 dx \frac{dy}{|y|^{2n+\epsilon}} \right]^{\frac{1}{2}} = A_{n\epsilon} B \, q^{-k\epsilon/2}.$$

Lemma (1.8). *If* m *satisfies the hypotheses of* (1.1) *then there is a constant* $A_{n\epsilon} > 0$ *independent of* $k, t \in \mathbb{Z}$ *such that*

$$\sup_{|\eta| \leq q^t} \int_{|\xi| > q^t} |(m_k)^{\vee}(\xi + \eta) - (m_k)^{\vee}(\xi)| d\xi$$

$$\leq A_{\epsilon n} (B + \|m\|_\infty) .$$

<u>Proof</u>. If $\epsilon > 0$ is as in (1.1) then

$$\int_{|y|\geq q^{\ell}} \int_{|x|=q^{\ell}} |m(x+y)-m(x)|^2 dx\, |y|^{-(2n+\epsilon)} dy$$

$$\leq 4(1-q^{-n})q^{\ell n} \|m\|_\infty^2 \int_{|y|\geq q^{\ell}} |y|^{-(2n+\epsilon)} dy$$

$$\leq A_{\epsilon n}\|m\|_\infty^2 \, q^{\ell n} \, q^{-\ell(n+\epsilon)} = A_{\epsilon n}\|m\|_\infty^2 \, q^{-\ell\epsilon} .$$

Let $m^{\ell}(x) = m(x)$ if $|x| = q^{\ell}$ and be zero otherwise. Then the estimate above holds as well with m replaced by m^{ℓ}, (1.2) does not change, so combining the estimates we get

$$\iint |m^{\ell}(x+y) - m^{\ell}(x)|^2 dx \frac{dy}{|y|^{2n+\epsilon}} \leq (B^2 + \|m\|_\infty^2) A_{n\epsilon} \, q^{-\ell\epsilon} .$$

From (1.7) it follows that,

$$\int_{|\xi|\geq q^k} |(m^{\ell})^{\vee}(\xi)| d\xi \leq A_{n\epsilon}(B + \|m\|_\infty) q^{-(\ell+k)\epsilon/2} .$$

m^s is supported on $\mathfrak{P}^{-s}$, so $(m^s)^{\vee}$ is constant on cosets of $\mathfrak{P}^s$, so $(m^s)^{\vee}(\xi+\eta) = (m^s)^{\vee}(\xi)$, whenever $|\eta| \leq q^{-s}$.

Suppose $|\eta| \leq q^t$. Note that $m_k = \sum_{-\infty}^{-k} m^s$. Then,

$$|(m_k)^{\vee}(\xi+\eta) - (m_k)^{\vee}(\xi)| \leq \sum_{s=-t+1}^{\infty} |(m^s)^{\vee}(\xi+\eta) - (m^s)^{\vee}(\xi)| .$$

If $|\xi| > q^t$, $|\eta| \leq q^t$. $|\xi+\eta| = |\xi|$. Thus,

$$\int_{|\xi|>q^t} |(m_k)^\vee(\xi+\eta) - (m_k)^\vee(\xi)|\,d\xi$$

$$\leq 2\sum_{s=-t+1}^{\infty} \int_{|\xi|>q^t} |(m^s)^\vee(\xi)|\,d\xi$$

$$\leq 2\, A_{\epsilon n}(B + \|m\|_\infty) \sum_{s=-t+1}^{\infty} q^{-(t+s)\epsilon/2}$$

$= A_{\epsilon n}(B + \|m\|_\infty)$, where the definition of $A_{\epsilon n}$ changes as needed.

Lemma (1.9). If m is as in (1.1) there is a constant $A > 0$, which depends only on ϵ, B, n and $\|m\|_\infty$, and in particular is independent of $\varphi \in \mathcal{S}$, $k \in \mathbb{Z}$ and $s > 0$ such that
$|\{x : |((m_k)^\vee * \varphi)(x)| > s\}| \leq A\|\varphi\|_1\, s^{-1}$.

Proof. Fix $s > 0$, $k \in \mathbb{Z}$, $\varphi \in \mathcal{S}$. Let $\varphi = \varphi_1^s + \varphi_2^s$ as in III (7.6) - (7.9). Then $\|\varphi_2^s\|_2^2 \leq q^n\|\varphi\|_1\, s$, and as usual we obtain,

$$|\{x : |(m_k)^\vee * \varphi_2^s(x)| > s/2\}| \leq 4q^n\|m\|_\infty^2\, \|\varphi\|_1\, s^{-1}.$$

We need a similar estimate for φ_1^s and we are done.

We now examine $(m_k)^\vee * \varphi_1^s(x)$. Recall that φ_1^s is supported on a set D_s which is the disjoint union, $\cup_t\, \omega_t$, ω_t a sphere,

$\int_{\omega_t} \varphi_1^s(x)\,dx = 0$, $\|\varphi_1^s\|_1 \le 2\|\varphi\|_1$, $|D_s| \le \|\varphi\|_1\, s^{-1}$. Let $y_t \in \omega_t$ for each t

$$(m_k)^\vee * \varphi_1^s(x) = \int \varphi_1^s(y)\,(m_k)^\vee(x-y)\,dy$$

$$= \sum_t \int_{y\in\omega_t} \varphi_1^s(y)\,(m_k)^\vee(x-y)\,dy$$

$$= \sum_t \int_{y\in\omega_t} \varphi_1^s(y)\,((m_k)^\vee(x-y) - (m_k)^\vee(x-y_t))\,dy .$$

Thus, $\displaystyle\int_{x\notin D_s} |(m_k)^\vee * \varphi_1^s(x)|\,dx$

$$\le \sum_t \int_{y\in\omega_t} \int_{x\notin\omega_t} |(m_k)^\vee(x-y) - (m_k)^\vee(x-y_t)|\,dx\,|\varphi_1^s(y)|\,dy$$

$$\le A_{\epsilon n}(B + \|m\|_\infty) \sum_t \int_{y\in\omega_t} |\varphi_1^s(y)|\,dy \qquad \text{(by (1.8))}$$

$$= A_{\epsilon n}(B + \|m\|_\infty)\|\varphi_1^s\|_1 \le A\|\varphi\|_1 ,$$

where A depends only on ϵ, n, B and $\|m\|_\infty$. Thus,

$|\{x : |(m_k)^\vee * \varphi_1^s(x)| > s/2\}|$

$$\le |D_s| + |\{x \notin D_s : |(m_k)^\vee * \varphi_1^s(x)| > s/2\}|$$

$$\le \|\varphi\|_1\, s^{-1} + \left[\int_{x\notin D_s} |(m_k)^\vee * \varphi_1^s(x)|\,dx\right] s^{-1}$$

$\le A\|\varphi\|_1\, s^{-1}$, and the proof of (1.9) is complete.

This also establishes (1.1).

2. Special cases of the multiplier theorem

We noted that if m is bounded and radial then m satisfies (1.1) with $B = 0$. More generally if $m(x+y) = m(x)$ whenever $|y| < |x|$ we get the same result. This condition can be written as follows: m is constant on cosets of $\mathfrak{P}^{k+1}$ in $\mathfrak{P}^k \approx \mathfrak{P}^{k+1}$ for all k.

We may generalize this condition as follows: Let s be a non-negative integer. Then we require that m be constant on cosets of $\mathfrak{P}^{k+s+1}$ in $\mathfrak{P}^k \approx \mathfrak{P}^{k+1}$ for all k, or equivalently, $m(x+y) = m(x)$ whenever $|y| < q^{-s}|x|$. For such an $m \in L^\infty$ we obtain,

$$\int_{|y|< q^{\ell}} \int_{|x|= q^{\ell}} |m(x+y) - m(x)|^2 dx\, |y|^{-(2n+\epsilon)} dy$$

$$\leq 4\, \|m\|_\infty^2 \,(1-q^{-n})^2 \Big(\sum_1^{s+1} q^{k(n+\epsilon)}\Big) q^{-\epsilon\ell} \,.$$

Thus, the conditions of (1.1) hold for any such m and any $\epsilon > 0$. However, a proof along these lines for this situation is unnecessarily deep. We will sketch a more elementary proof for this special, but important case.

<u>Lemma (2.1)</u>. <u>If</u> $m \in L^\infty$ <u>and</u> $m(x+y) = m(x)$ <u>whenever</u> $|y| < q^{-s}|x|$, s <u>a non-negative integer, then for all</u> $k \in \mathbb{Z}$ $(m_k)^\vee(x+y) = (m_k)^\vee(x)$ <u>whenever</u> $|y| < q^{-s}|x|$.

The proof is left as a modest exercise.

<u>Theorem (2.2).</u> <u>If</u> $m \in L^{\infty}$ <u>and</u> $m(x+y) = m(x)$ <u>whenever</u> $|y| < q^{-r}|x|$, r <u>a non-negative integer, then the conclusion of (1.1) holds with</u> A <u>depending only on</u> r, n <u>and</u> $\|m\|_{\infty}$.

<u>Proof</u>. Let $D_s = \bigcup_t \omega_t$, $\omega_t = y_t + \mathfrak{P}^{\ell_t}$ as in the proof of (1.9). Then let $D_s^* = \bigcup_t (\omega_t)^*$ where $(\omega_t)^* = y_t + \mathfrak{P}^{\ell_t - r}$. Then $|D_s^*| \leq q^{nr}|D_s| \leq A_{nr}\|\varphi\|_1 \; s^{-1}$.

As in (1.9) we have

$$(m_k)^{\vee} * \varphi_1^s(x) = \sum_t \int_{y \in \omega_t} \varphi_1^s(y)\,(m_k)^{\vee}(x-y)\,dy .$$

Let $y_t \in \omega_t$ and suppose $x \notin (D_s)^*$, so $x \notin (\omega_t)^*$. Then $x-y = (x-y_t) + (y-y_t)$. $|y-y_t| \leq q^{-\ell_t} = q^{-\ell_t+r} \cdot q^{-r} < |x-y|\,q^{-r}$, when $y \in \omega_t$. Thus, $(m_k)^{\vee}(x-y)$ takes the constant value $(m_k)^{\vee}(x-y_t)$ on ω_t when $x \notin D_s^*$, so $(m_k)^{\vee} * \varphi_1^s(x) = 0$ if $x \notin D_s^*$ since $\int_{\omega_t} \varphi_1^s(y)\,dy = 0$.

Thus, $|\{x : |(m_k)^{\vee} * \varphi_1^s(x)| > s/2\}| \leq |D_s^*| \leq A_{nr}\|\varphi\|_1 \; s^{-1}$.

The rest of the proof goes as before and the entire family of lemmas (1.5) through (1.8) and much of the argument of (1.9) is avoided.

Example. If π is a unitary character on K^* that is ramified of degree h, then π satisfies the conditions of (2.2) with $\|\pi\|_\infty = 1$ and $r = h-1$. If π is unitary and unramified, then π is radial and the conditions are satisfied with $r = 0$.

3. Applications of the multiplier theorem to Fourier series

We recall some facts about Fourier series from II §6.

$\{u(n)\}_{n=0}^\infty$ is a complete set of coset representatives of $\mathfrak{O}$ in K. If $\chi_n \equiv \chi_{u(n)}|_{\mathfrak{O}}$ then $\{\chi_n\}_{n=0}^\infty$ is a complete set of characters on $\mathfrak{O}$. We also take the point of view that χ_n is defined on K, but is supported on $\mathfrak{O}$.

$u(n) = 0$ iff $n = 0$ and $|u(n)| = q^{k+1}$ when $q^k \le n < q^{k+1}$. If $n = r\cdot q^k+s$, $r \ge 0$, $0 \le s < q^k$, then $\chi_n = \chi_{rq^k}\chi_s$. Observe that for $k = 0,1,\ldots$, $\{u(r\cdot q^k)\}_{r=0}^{q-1}$ are q distinct coset representatives of $\mathfrak{P}^{-k}$ in $\mathfrak{P}^{-(k+1)}$. To see this, note that $u(rq^k) = \mathfrak{p}^{-k}u(r)$ and verify for $k = 0$.

The Dirichlet kernel is $D_n(x) = \sum_{\nu=0}^{n-1} \chi_\nu(x)$, $n \ge 1$.

For $f \in L^1$, let $\hat{f}(n) = \int_{\mathfrak{O}} f(x)\bar{\chi}_n(x)\,dx$. Then we write $f(x) \approx \sum \hat{f}(n)\chi_n(x)$, and we let $S_n f(x) = \sum_{\nu=0}^{n-1} \hat{f}(\nu)\chi_\nu(x)$, and we recall that $S_n f = D_n * f$ where the convolution is on $\mathfrak{O}$, or is on K if we think of the functions as supported on $\mathfrak{O}$.

We will study the L^p properties of the maps $f \to S_n f$, $1 < p < \infty$. In the real case the relationships $S_n^* f(x) = \sin nx(\cos n(\cdot)f)^{\approx}(x) - \cos nx(\sin n(\cdot)f)^{\approx}(x)$, where S_n^* represents "modified" partial sums and $(\ )^{\approx}$ is the conjugate transform, are used in that case. We proceed with the development of analogous relations. Each T_n, defined below is a "conjugate transform".

<u>Definition.</u> For $n \geq 1$, $D_n^{\#} = \overline{\chi}_n D_n$, $T_n f = D_n^{\#} * f$, $f \in L^1$.

<u>Proposition (3.1)</u>. For $n \geq 1$, $f \in L^1$. $S_n f = \chi_n(T_n(\overline{\chi}_n f))$.

<u>Proof</u>. $S_n f(x) = \int_{\mathfrak{O}} f(z) D_n(x-z)dz$

$$= \int f(z)\chi_n(x-z)D_n^{\#}(x-z)dz$$

$$= \chi_n(x) \int (f\overline{\chi}_n)(z)D_n^{\#}(x-z)dz = \chi_n(T_n(\overline{\chi}_n f)) .$$

<u>Proposition (3.2)</u>. (a) $(\Phi_0 D_n^{\#})^\wedge(x) = 0$ or 1 .

(b) If $|y| < |x|$ then $(\Phi_0 D_n^{\#})^\wedge(x+y) = (\Phi_0 D_n^{\#})^\wedge(x)$.

Proof. For this proof we take the point of view that the functions are defined on K, but supported on $\mathfrak{O}$.

$\Phi_0 D_n^{\#} = \sum_{\nu=0}^{n-1} \Phi_0\, \chi_{(u(\nu)-u(n))}$ and $\{u(\nu)-u(n)\}_{\nu=0}^{n-1}$ are n distinct coset representatives of $\mathfrak{O}$. Since $(\Phi_0\, \chi_h)^{\wedge} = \tau_h \Phi_0$ we see that $(\Phi_0 D_n^{\#})^{\wedge}$ is the characteristic function of n distinct cosets of $\mathfrak{O}$ and hence is equal to zero or one. That takes care of (a).

To establish (b) we need to show that these n cosets fill out cosets of $\mathfrak{P}^{-\ell}$ in $\mathfrak{P}^{-\ell-1} \sim \mathfrak{P}^{-\ell}$, for $\ell = 0,1,\ldots,k$, $1 \le n < q^{k+1}$ and are contained in $\mathfrak{P}^{-k-1}$. We proceed by induction on k.

If $k = 0$, then $1 \le n < q$, $\bar{\chi}_n D_n = \sum_{\nu=0}^{n-1} \chi_{(u(\nu)-u(n))}$. The $\{u(\nu)-u(n)\}_{\nu=0}^{n-1}$ are n distinct coset representatives of $\mathfrak{O} = \mathfrak{P}^0$ in $\mathfrak{P}^{-1} \sim \mathfrak{P}^0$ and so our result holds if $k = 0$. (Notice the only way this could fail is if $u(\nu)-u(n) \in \mathfrak{O}$.)

We now assume that the result is established for $k-1$, $k \ge 1$, k fixed and show that it holds for k.

We may assume that $q^k \le n < q^{k+1}$ and write $n = r \cdot q^k + s$, $1 \le r < q$, $0 \le s < q^k$. Then

$$\bar{\chi}_n D_n = (\bar{\chi}_{r\cdot q^k}\, \bar{\chi}_s)\left[\sum_{\rho=0}^{r-1} \chi_{\rho\cdot q^k} \sum_{\sigma=0}^{q^k-1} \chi_\sigma + \chi_{r\cdot q^k} \sum_{\sigma=0}^{s-1} \chi_\sigma\right]$$

$$= \left[\sum_{\ell=0}^{r-1} (\bar{\chi}_{r\cdot q^k}\, \chi_{\rho\cdot q^k})\right]\left[\sum_{\sigma=0}^{q^k-1} (\bar{\chi}_s \chi_\sigma)\right] + \bar{\chi}_s D_s \quad .$$

By the induction hypothesis, the last term satisfies the requirements of the proposition and the corresponding cosets are all contained in $\mathfrak{B}^{-k}$.

In the sum $\sum_{\sigma=0}^{q^k-1} \overline{\chi}_s \chi_\sigma$ we see that $\{u(\sigma)-u(s)\}_{\sigma=0}^{q^k-1}$ are q^k distinct cosets of $\mathfrak{L}$ in $\mathfrak{B}^{-k}$ so the sum is $D_{q^k} = \sum_{\sigma=0}^{q^k-1} \chi_\sigma$. In the sum $\sum_{\rho=0}^{r-1} \overline{\chi}_{r\cdot q^k} \chi_{\rho\cdot q^k}$, the terms correspond to the r coset representatives, $\{u(\rho q^k)-u(rq^k)\}_{\rho=0}^{r-1}$ are r distinct coset representatives of $\mathfrak{B}^{-k}$ in $\mathfrak{B}^{-k-1} \sim \mathfrak{B}^{-k}$ and since each term multiplies D_{q^k} we fill out those complete cosets and the result is established.

<u>Corollary (3.3)</u>. <u>If</u> $n \geq 1$, $|y| < |x| \leq 1$ <u>then</u> $D_n^{\#}(x+y) = D_n^{\#}(x)$.

<u>Proof</u>. This is an immediate consequence of (3.2) and (2.1).

<u>Theorem (3.4)</u>. <u>If</u> $f \in L^p(\mathfrak{L})$, <u>then</u> $\|S_n f\|_p \leq A_p \|f\|_p$, $\|T_n f\|_p \leq A_p \|f\|_p$ <u>where</u> $A_p > 0$ <u>is independent of</u> f <u>and</u> $A_p = O(p^2/(p-1))$ <u>as</u> $p \to 1$ <u>and</u> $p \to +\infty$, $1 < p < \infty$.

<u>Proof</u>. $S_n f = D_n * f$ and $T_n f = D_n^{\#} * f$, D_n and $D_n^{\#}$ are in L^1 so it will suffice to assume that $f \in \mathcal{S}$. Furthermore, from (3.1), we

see that if the result holds for T_n , then it holds for S_n with the same constant. Thus:

$$\|S_n f\|_p = \|\chi_n(T_n(\bar{\chi}_n f))\|_p = \|T_n(\bar{\chi}_n f)\|_p$$

$$\leq A_p \|(\bar{\chi}_n f)\|_p = A_p \|f\|_p \ .$$

So we check to see if $(D_n^{\#} \Phi_0)^\wedge$ is a multiplier on L^p. But the result of (3.2) shows that $(D_n^{\#} \Phi_0)^\wedge$ satisfies the requirements of (2.2) with $\|m\|_\infty = 1$ and $r = 0$.

__Corollary (3.5)__. __If__ $f \in L^p$, $1 < p < \infty$ __then__ $S_n f \to f$ __in__ L^p __as__ $n \to \infty$.

__Proof__. Fix $\varepsilon > 0$. Write $f = b + g$, where $b \in L^p$, $\|b\|_p < \varepsilon/(1 + A_p)$, and $g \in \mathcal{S}(\mathfrak{L})$ where A_p is the constant of (3.4). For n large enough $S_n g = g$, so if n is large enough $\|S_n f - f\|_p \leq \|S_n b\|_p + \|b\|_p \leq A_p\|b\|_p + \|b\|_p = (1 + A_p)\|b\|_p < \varepsilon$.

__Corollary (3.6)__. __If__ $f \in L^p(\mathfrak{L})$, $1 < p < \infty$, __let__

$$T_n^* f(x) = \sup_{k \geq 0} \left| \int_{|x-t| \geq q^{-k}} f(t) D_n^{\#}(x-t)\, dt \right| .$$

__Then there are constants__ $C_p > 0$, __independent of__ f, n __and__ $s > 0$ __such that__,

$$|\{x : T_n^* f(x) > s\}| \leq (C_p \|f\|_p \, s^{-1})^p \quad \text{\underline{and}}$$

$$C_p = O(p^3/(p-1)^2) \quad \text{\underline{as}} \quad p \to 1 \quad \text{\underline{or}} \quad p \to \infty.$$

<u>Proof.</u> Fix f and n. $T_n^* f(x)$ exists as a measurable function and is finite for all $x \in \Omega$, since $|D_n^{\#}| \le n$ and $f \in L^1$ so $T_n^* f \in L^\infty$ and is bounded by $n\|f\|_1$.

For each $x \in \Omega$ let $k(x) \in \mathbb{Z}$ so that

$$T_n^* f(x) < 2\left| \int_{|x-t| > q^{-k(x)}} f(t) D_n^{\#}(x-t)\,dt \right| .$$

Suppose $z \in x + \mathfrak{P}^{k(x)}$. Then if $|x-t| \ge q^{-k(x)}$ we see that $D_n^{\#}(x-t) = D_n^{\#}(z-t)$ (since $|x-z| < |x-t|$, using (3.3)). Thus, if $z \in x + \mathfrak{P}^{k(x)}$,

$$(1/2) T_n^* f(x) \le \left| \int_{|x-t| > q^{-k}} f(t) D_n^{\#}(z-t)\,dt \right|$$

$$\le \left| \int [\tau_x \Phi_{k(x)}(t) f(t)] D_n^{\#}(z-t)\,dt \right| + \left| \int f(t) D_n^{\#}(z-t)\,dt \right|$$

$$= |T_n(\tau_x \Phi_{k(x)} f)(z)| + |T_n f(z)| .$$

We fix x and note that both functions on the right hand side are continuous in z. We average over $x + \mathfrak{P}^{k(x)}$ and get,

$$(1/2) T_n^* f(x) \le q^{k(x)} \int_{|x-z| \le q^{-k(x)}} |T_n(\tau_x \Phi_{k(x)} f)(z)|\,dz + M(T_n f)(x) .$$

Let A_p be the norm of T_p as an operator on L^p, $1 < p < \infty$, B_p the norm of M as an operator on L^p, $1 < p \le \infty$ and B_1 the weak type (1,1) "norm" of M. Then $A_p = O(p^2/(p-1))$ as $p \to 1$ or $p \to \infty$, $B_p = O(p/(p-1))$ as $p \to 1$ or $p \to \infty$.

$$q^{k(x)} \int_{|x-z| \le q^{-k(x)}} |T_n(\tau_x \Phi_{k(x)} f)(z)|\,dz$$

$$\le q^{k(x)/p} \Big[\int_{|x-z| \le q^{-k(x)}} |T_n(\tau_x \Phi_{k(x)} f)(z)|^p dz \Big]^{1/p}$$

$$\le A_p\, q^{k(x)/p} \Big[\int |(\tau_x \Phi_{k(x)} f)(z)|^p dz \Big]^{1/p}$$

$$= A_p \Big[q^{k}(x) \int_{z \in x + \mathfrak{P}^{k(x)}} |f(z)|^p dz \Big]^{1/p}$$

$$\le A_p \Big[M(|f|^p)(x) \Big]^{1/p} .$$

Thus, $(1/2)T_n^* f(x) \le A_p \Big[M(|f|^p)(x) \Big]^{1/p} + M(T_n f)(x)$,

$$|\{x : |T_n^* f(x)| > s\}| \le |\{x : M(|f|^p)(x) > (s/4\,A_p)^p\}| + |\{x : M(T_n f)(x) > s/4\}|$$

$$\le B_1 \|f\|_p^p (4A_p\, s^{-1})^p + (4\|M(T_n f)\|_p\, s^{-1})^p$$

$$\le ((B_1^{1/p} 4A_p)\|f\|_p\, s^{-1})^p + (4A_p B_p \|f\|_p\, s^{-1})^p$$

$$= \Big[(4A_p (B_1 + B_p^p)^{1/p}) \|f\|_p\, s^{-1} \Big]^p .$$

<u>Corollary (3.7)</u>. <u>Let</u> ω, ω' <u>represent spheres in</u> $\mathfrak{O}$. <u>Suppose</u> $f \in L^\infty(\omega)$. <u>Then for</u> $x \in \omega$ <u>let</u>

$$T_n^*(x;\omega) = \sup_{x\in\omega'\subset\omega}\left|\int_{\omega\sim\omega'} f(t)D_n^{\#}(x-t)dt\right| .$$

Then $|E_s| = |\{x \in \omega : T_n^*(x;\omega) > s\}|$

$$\leq e^2[\exp(-s/Ne\|f\|_{\infty,\omega})]|\omega| ,$$

where $N > 0$ _is independent of_ ω, f _and_ $s > 0$.

Proof. We may assume $\|f\|_{\infty;\omega} = 1$. Let $g = \begin{cases} f, & x \in \omega \\ 0, & x \notin \omega \end{cases}$.

Then $T_n^*(x;\omega) \leq 2\, T_n^* g(x)$. For $p \geq 2$, we have

$$|E_s| = |\{x : T_n^*(x;\omega) > s\}| \leq |\{x : T_n^* g(x) > s/2\}|$$

$$\leq (2\, C_p \|g\|_p\, s^{-1})^p \leq (2\, C_p\, s^{-1}\|f\|_{\infty,\omega})^p|\omega|$$

$$\leq [N\, p\, s^{-1}]^p\, |\omega| ,$$

where $N > 0$ is independent of f, p, ω and $s > 0$.

We look for the minimum of $[N\, p\, s^{-1}]^p$, $2 \leq p < \infty$.

Consider an expression of the form $F(p) = (A_p)^p$, $A > 0$. It takes its minimum $\exp(-1/eA)$ at $p = 1/eA$. Thus, if $s/eN \geq 2$, we have

$$|E_s| \leq [\exp(-s/Ne)]|\omega| \leq e^2[\exp(-s/Ne)]|\omega| .$$

Otherwise, $|E_s| \leq |\omega| = e^2\, e^{-2}\, |\omega|$

$$\leq [\exp(-s/NE)]\, |\omega| .$$

4. Singular integral operators

We indicated at the end of §2 that an interesting class of multipliers on L^p, $1 < p < \infty$ is obtained by letting $m(x) = \pi(x)$, π a unitary multiplicative character. If π is the identity then the multiplier map is convolution with the dirac delta. It was observed in II §5 (5.7) that $((1/\Gamma(\pi^{-1}))(\pi|\cdot|)^{-1})^\wedge = \pi$. It is implicit in that development, and the material in III §4 that if π is ramified, and we set $K(x) = (1/\Gamma(\pi^{-1}))\pi^{-1}(x)|x|^{-1}$, then

$$((PK)*\varphi)(x) = \lim_{k\to-\infty} \int_{|z|\geq q^k} \varphi(x-z)K(z)dz$$ is well defined for all

$\varphi \in \mathcal{S}$ and $((PK)*\varphi)^\wedge(x) = \pi(x)\hat{\varphi}(x)$. Hence, such operators are bounded on $L^p \cap \mathcal{S}$ to L^p, $1 < p < \infty$.

We state here, an elementary version of a generalization of this fact in the setting of operators on $L^p(K^n)$. We omit the proofs.

<u>Theorem (4.1)</u>. <u>Suppose</u> $\Omega \in L^\infty(K^n)$, $\Omega(\mathfrak{p}^k x) = \Omega(x)$ <u>for all</u> $k \in \mathbb{Z}$ <u>and</u> $\int_{|x|=1} \Omega(x)dx = 0$.

(a) <u>If</u>

$$\text{(4.2)} \qquad \sup_{|y|=1} \sum_{j=1}^{\infty} \int_{|x|=1} |\Omega(x+\mathfrak{p}^j y) - \Omega(x)|dx < \infty ,$$

<u>then for</u> $1 < p < \infty$ <u>there is a constant</u> $A_p > 0$ <u>independent of</u> f

<u>and</u> $k \in \mathbb{Z}$ <u>such that if</u>

$$\tilde{f}_k(x) = \int_{|z|>q^k} f(x-z)\frac{\Omega(z)}{|z|^n}\,dz \quad , \quad \underline{\text{then}}$$

$$\tilde{f}_k \in L^p,\ \|\tilde{f}_k\|_p \le A_p\|f\|_p \quad \text{for all} \quad f \in L^p,\ k \in \mathbb{Z}\,.$$

<u>Furthermore,</u> $\tilde{f} = \lim_{k\to-\infty} \tilde{f}_k$ <u>exists in the</u> L^p<u>-norm and</u>

$\|\tilde{f}\|_p \le A_p\|f\|_p$.

<u>If</u> $f \in L^1$, <u>then there is a constant</u> $A_1 > 0$, <u>independent of</u> $f \in L^1$, $k \in \mathbb{Z}$ <u>and</u> $s > 0$ <u>such that</u>

$$|\{x : |\tilde{f}_k(x)| > s\}| \le A_1\|f\|_1\, s^{-1} ,$$

$\tilde{f}_k \to \tilde{f}$ <u>exists in measure as</u> $k \to -\infty$, <u>and</u> $|\{x:|\tilde{f}(x)|>s\}| \le A_1\|f\|_1 s^{-1}$

(b) <u>If, in addition to (4.2) we have</u>

$$(4.3) \qquad \sum_{j=1}^{\infty} \sup_{|x|=1} \Big|\int_{|y|\le 1} (\Omega(x+\rho^j y) - \Omega(x))\,dy\Big| < \infty ,$$

<u>then for</u> $1 \le p < \infty$ $\tilde{f}_k(x)$ <u>converges to</u> $\tilde{f}(x)$ a.e. <u>if</u> $f \in L^p$.

<u>Furthermore, there are constants</u> $B_p > 0$ <u>independent of</u> $f \in L^p$, $1 \le p < \infty$ <u>such that if</u>

$$(\tilde{f})^*(x) = \sup_k |f_k(x)| \quad \text{then}$$

$$(\tilde{f})^* \in L^p,\ \|(\tilde{f})^*\|_p \le B_p\|f\|_p\ ,\ 1 < p < \infty,\ \underline{\text{and}}$$

<u>for each</u> $s > 0$, $|\{x:(\tilde{f})^*(x) > s\}| \le B_1\|f\|_1\, s^{-1}$.

Remarks. It is an easy exercise to see that (4.2) immediately implies (1.8) for $\Omega(x)/|x|^n$. We also observe that (4.3) is a weakish sort of Dini condition. Note that if Ω satisfies a Hölder continuity condition on $\mathfrak{D}^*$, or is merely Dini continuous, then both (4.2) and (4.3) are satisfied.

Observe also that since $\Omega \in L^\infty(\mathfrak{D}^*)$ and $\int_{\mathfrak{D}^*} \Omega(x)dx = 0$ that

$P\,\Omega(\cdot)|\cdot|^{-n} \in \mathcal{S}'$ by an easy calculation. In fact if $\Omega \in L^1(\mathfrak{D}^*)$ we would still have that $P\,\Omega(\cdot)|\cdot|^{-n} \in \mathcal{S}'$. With some modest complication in the statement of the theorem and in its proof, the conclusions remain valid.

Operators of the form $f \to \tilde{f} = \lim_k \tilde{f}_k$ where the $\tilde{f}_k$ are defined by removing small "area" about a singularity in a "natural" integral definition of $\tilde{f}$ are known as singular integral operators.

5. Notes for Chapter VI

A general reference for §1 - §3 is Taibleson [5]; for §4 see Phillips and Taibleson [1].

§1. This entire section is devoted to the proof of (1.1) which is the local field variant of Hörmander's version of the Mihlin multiplier theorem (Hörmander [1, Thm.2.5]). To show that this is indeed true we will explain a little.

Hörmander's theorem states that if $m \in L^{\infty}(R^n)$, and if the restriction of m to the set $\{R/2 \leq |x| \leq 2R\}$, for each $R > 0$, is in the Lebesgue potential space $L^2_{[\frac{n}{2}]+1}(R^n)$ with norm bounded by $BR^{-([\frac{n}{2}]+1-\frac{n}{2})}$ (some $B > 0$), then m is a multiplier on $L^p(R^n)$, $1 < p < \infty$.

Our theorem requires that $m \in L^{\infty}(K^n)$ and that for each $\ell \in \mathbb{Z}$, the restriction of m to the set $\{|x| = q^{\ell}\}$ is in the Lipschitz space $\Lambda^{2,2}_{\frac{n}{2}+\frac{\epsilon}{2}}(K^n)$ (use the norms defined in IV §2 to extend the definition of Λ^p_{α} of III §8 in the obvious way) with norm bounded by $B\, q^{-\epsilon\ell/2}$ (some $B > 0$, $\epsilon > 0$, and all $\ell \in \mathbb{Z}$), then m is a multiplier on $L^p(K^n)$, $1 < p < \infty$. Since $\Lambda^{2,2}_{\alpha} = L^2_{\alpha}$ (a standard fact for R^n, an easy consequence of (1.6) for local fields), we see that the two statements are essentially identical, except that for local fields we have a formulation for any $\epsilon > 0$; while for R^n we have it for the specific value, $\epsilon/2 = [n/2] + 1 - n/2$.

§3. The modified Dirichlet kernels, $D_n^{\#}$, studied in this section, were first introduced by Paley [1], in his study of Fourier series on the dyadic group, 2^{ω}. He used it (implicitly) to prove his Theorem VI, which is a special case of our Theorem (3.4), and in a manner that is strictly analogous to our proof.

The result, (3.6), on the maximal version of the operators $\{T_n\}$, reappears in Chapter VIII as (2.9). It is a crucial step in the proof that the Fourier series of a function in L^p, for any $p > 1$, converges a.e. to the function.

In the proof of (3.6) we stated that the norm, B_p, of the Hardy-Littlewood maximal operator satisfies the condition: $B_p = O(1)$ as $p \to \infty$. However, in proving that the maximal operator was bounded for $1 < p < \infty$, we used the Marcinkiewicz interpolation theorem (IV(1.7)) and the best that result yields is $B_p = O(p)$ as $p \to \infty$. The problem is, that $f \to Mf$ is sub-linear, otherwise we could use the Riesz-Thorin interpolation theorem and from the fact that $\|Mf\|_2 \leq B_2 \|f\|_2$, $\|Mf\|_\infty \leq \|f\|_\infty$, we would have, $B_p \leq B_2^{2/p} = O(1)$ as $p \to \infty$. We get our conclusion by linearizing the maximal operator. Note that $\|Mf\|_p = \|\sup_k |f(x,k)|\|_{p,dx}$, so that $\|Mf\|_p$ is the norm of $f(x,k)$ on the mixed norm space $L^{\infty,p}(\mathbb{Z},K^n)$, and $f(\cdot) \to f(\cdot,k)$ is linear. Our result then follows from the Riesz-Throin theorem for mixed norm spaces, since the map is bounded from $L^2(K^n)$ to $L^{\infty,2}(\mathbb{Z},K^n)$ with norm B_2, and from $L^\infty(K^n)$ to $L^{\infty,\infty}(\mathbb{Z},K^n)$ with norm 1, and so from $L^p(K^n)$ to $L^{\infty,p}(\mathbb{Z}, K^n)$ with the norm as advertised. See Benedek and Panzone [1,p.316] for details.

§4. The results in this section are the local field variants of the wo in singular integral operator theory of Calderón and Zygmund [1]. For $n = 1$ and K a p-adic or p-series field, this extension was first given by Phillips [1] for the case $\Omega(x+y) = \Omega(x)$, $|y| < q^{-s}|x|$, s a positiv integer.

The results of this section for Ω a multiplicative character, will be used in the next chapter.

Chapter VII. Conjugate systems of regular functions and an F. and M. Riesz theorem

1. A fundamental lemma

Lemma (1.1). *Let* $F(x,k) = (f_0(x,k), f_1(x,k), \cdots, f_m(x,k))$ *be a vector-valued function with each component* $f_j(x,k)$, $j = 0,1,\ldots,m$ *being regular on* $K^n \times \mathbb{Z}$. *Suppose there is a* p_0, $0 < p_0 < 1$, *such that* $|F(x,k)|^{p_0}$ *is sub-regular, with*

$|F(x,k)| = \left[\sum_{j=0}^{m} |f_j(x,k)|^2\right]^{1/2}$. *Suppose further that for some*

$p > p_0$, *and* $A > 0$,

(1.2) $\displaystyle\int_{K^n} |F(x,k)|^p \, dx \leq A < \infty$ *for all* $k \in \mathbb{Z}$,

then,

(a) $f_j(x) = \lim_{k \to -\infty} f_j(x,k)$ *exists* a.e., $j = 0,1,\ldots,m$.

(b) $\displaystyle\lim_{k \to -\infty} \int_{K^n} |F(x,k) - F(x)|^p \, dx = 0$, *with* $F(x) = (f_0(x), f_1(x), \ldots, f_m(x))$.

Moreover,

(c) $f_j^*(x) = \sup_{k \in \mathbb{Z}} |f_j(x,k)| \in L^p(K^n)$, $j = 0,1,\ldots,m$.

Proof. Let $p_1 = p/p_0 > 1$. For all $k \in \mathbb{Z}$,

$\displaystyle\int (|F(x,k)|^{p_0})^{p_1} \, dx = \int |F(x,k)|^p \, dx \leq A < \infty$.

Applying IV (3.7), and observing that $p_1 > 1$, and $|F(x,k)|^{p_0}$ is non-negative and sub-regular we see that $|F(x,k)|^{p_0}$ has a regular majorant $m(x,k)$, which is the regularization of $m \in L^{p_1}(K^n)$. Let $m^*(x) = \sup_k m(x,k)$ $(m(x,k) \geq 0)$. Then $m^* \in L^{p_1}(K^n)$ and in particular, $m^*(x) < \infty$ a.e.

Then for a.e. $x \in K^n$, $j = 0,1,\ldots,m$,

$$f_j^*(x) = \sup_k |f_j(x,k)| \leq \sup_k |F(x,k)|$$

$$\leq \sup_{k \in \mathbb{Z}} (m(x,k))^{1/p_0} = (m^*(x))^{p_1/p} < \infty .$$

Also, $\int |f_j^*(x)|^p \, dx \leq \int (m^*(x))^{p_1} < \infty$, so $f_j \in L^p$, $j = 0,1,\ldots,m$. This establishes (c).

Since $f_j(x) < \infty$ a.e., $f_j(x,k)$ converges a.e. to a function $f_j(x)$ as $k \to -\infty$ (V (2.3)). This establishes (a).

Thus, $F(x) = \lim_{k \to -\infty} F(x,k)$ exists a.e.,

$$\lim_{k \to -\infty} (F(x,k) - F(x)) = 0, \text{ a.e.,}$$

$$|F(x,k)-F(x)|^p \leq 2^p(|F(x,k)|^p + |F(x)|^p) \leq 2^{p+1}(m^*(x))^{p_1} \in L^1 .$$

By the dominated convergence theorem,

$$\lim_{k \to -\infty} \int_{K^n} |F(x,k)-F(x)|^p \, dx = 0.$$

This establishes (b) and the proof is complete.

2. Construction of conjugate systems on $K \times \mathbb{Z}$

We review some basic structural facts about K. Let $\mathfrak{O}^* = \mathfrak{O} \sim \mathfrak{P}$ be the group of units in K^*. Then there is an element $\epsilon \in \mathfrak{O}^*$, ϵ of order $q-1$ such that

$\mathfrak{O}/\mathfrak{P} = \{0 + \mathfrak{P}, \epsilon + \mathfrak{P}, \epsilon^2 + \mathfrak{P}, \ldots, \epsilon^{q-1} + \mathfrak{P} = 1 + \mathfrak{P} = A\}$, $\mathfrak{O}/\mathfrak{P} \cong GF(q)$.

<u>N.B.</u> We assume that q is odd.

Let π be a multiplicative unitary character that is ramified of degree 1 and homogeneous of degree zero. That is, π is constant on cosets of $A = 1 + \mathfrak{P}$ and $\pi(\mathfrak{p}^k x) = \pi(x)$ for all $k \in \mathbb{Z}$. Then π is a function of $\{\epsilon^k\}_{k=1}^{q-1}$ and takes its values in the group of $(q-1)$-<u>st</u> roots of unity. Since $\pi(\epsilon^k) = (\pi(\epsilon))^k$, π is determined by $\pi(\epsilon)$, and $(\pi(\epsilon))^{q-1} = 1$.

Let π be fixed, with $\pi(\epsilon) = \zeta$, a primitive $(q-1)$-<u>st</u> root of unity. The collection $\{\pi^\ell\}_{\ell=0}^{q-2}$ is a cyclic group of order q-1, and consists of the unitary multiplicative characters that are homogeneous of degree zero and ramified of degree 1, together with the identity character.

$\pi^\ell(\epsilon^k) = \zeta^{k\ell}$, and $\pi(-x) = -\pi(x)$ for all $x \in K^*$.
That is, π is odd. This follows since, $-1 \in \epsilon^{(q-1)/2} + \mathfrak{P}$. Thus, $\pi(-1) = \pi(\epsilon^{(q-1)/2}) = \zeta^{(q-1)/2}$. But $(\pi(-1))^2 = \pi((-1)^2) = \pi(1) = 1$, so $\pi(-1) = \pm 1$. If $\pi(-1) = 1$, then $\zeta^{(q-1)/2} = 1$, which violates the requirement that the order of ζ is $(q-1)$.

We now define some conjugate kernels. Let

$$Q^{\ell}(x) = (1/\Gamma(\pi^{\ell}))\pi^{\ell}(x)|x|^{-1},\ x \neq 0,\ \ell = 1,2,\ldots,q-2.$$

We have seen that $P\,Q^{\ell} \in \mathscr{S}'$ and we identify Q^{ℓ} with the distribution $P\,Q^{\ell}$.

Let $Q^{\ell}(\cdot,k) = Q^{\ell} * R(\cdot,k)$. It is easy to see that

$$Q^{\ell}(x,k) = Q^{\ell}(x)(1-\Phi_{-k}(x)) = \begin{cases} Q^{\ell}(x), & |x| > q^{k} \\ 0 & , |x| \le q^{k} \end{cases}.$$

For a "nice" function f on K, define T_{ℓ} and T'_{ℓ} as follows:

$$T_{\ell}f = \lim_{k \to -\infty} Q^{\ell}(\cdot,k) * f$$

$$(T'_{\ell}f)^{\wedge} = \pi^{-\ell}\hat{f} \ .$$

Using the language of the last chapter (VI) we see that T_{ℓ} is a singular integral operator and T'_{ℓ} is a multiplier transform. Moreover, they agree on $\mathscr{S}$ and so have identical extensions to L^{p}, $1 < p < \infty$. Furthermore, Let $Q^{0}(x,k) = R(x,k)$. Then let $T_{0}f = \lim_{k \to -\infty} Q^{0}(\cdot,k) * f = f$, and $(T'_{0}f)^{\wedge} = \pi^{0}\hat{f} = \hat{f}$, so that $T_{0} = T'_{0}$ and the collection $\{T_{\ell}\}_{\ell=0}^{q-2}$ is a cyclic group of operators on $\mathscr{S}$, $T_{\ell}T_{n} = T_{\ell+n}$, where we take the indices $\mathrm{mod}(q-1)$, and $(T_{\ell}f)^{\wedge} = \pi^{-\ell}\hat{f}$, at least for $f \in \mathscr{S}$.

<u>Notation</u>. Let $\{\varepsilon_m^j\}_{j=1}^{q-1}$, $m \in \mathbb{Z}$ be defined by $\varepsilon_m^j = \mathfrak{p}^{-(m+1)}\varepsilon^j$.

Then $\{\varepsilon_m^j\}$, for each m, is a complete set of coset representatives of $\mathfrak{P}^{-m}$ in $\mathfrak{P}^{-(m+1)} \approx \mathfrak{P}^{-m}$.

We will now compute $T_\ell f(x,k)$ explicitly for "nice enough" f, and $\ell \neq 0$. What we require is that $T_\ell f(\circ,k) = T_\ell f * R(\cdot,k) = f * Q^\ell(\cdot,k)$, makes sense.

$$(2.1) \qquad T_\ell f(x,k) = (Q^\ell(\cdot,k) * f)(x)$$

$$= \frac{1}{\Gamma(\pi^\ell)} \int_{|t|>q^k} f(x-t)\frac{\pi^\ell(t)}{|t|}\,dt$$

$$= \frac{1}{\Gamma(\pi^\ell)} \sum_{m=k}^{\infty} q^{-(m+1)} \int_{|t|=q^{m+1}} f(x-t)\pi^\ell(t)\,dt$$

$$= \frac{1}{\Gamma(\pi^\ell)} \sum_{m=k}^{\infty} q^{-(m+1)} \sum_{j=k}^{q-1} \pi^\ell(\varepsilon_m^j) \int_{\varepsilon_m^j+\mathfrak{P}^{-m}} f(x-t)\,dt$$

$$= \frac{1}{q\Gamma(\pi^\ell)} \sum_{m=k}^{\infty} \sum_{j=1}^{q-1} \pi^\ell(\varepsilon^j) f(x-\varepsilon_m^j,m)$$

Let $d_k f(x) = f(x,k) - f(x,k+1)$. Then,

$T_\ell(d_k f) = T_\ell f(\cdot,k) - T_\ell f(\cdot,k+1) = d_k(T_\ell f)$, and so

$$(2.2) \qquad T_\ell(d_k f)(x) = d_k(T_\ell f)(x)$$

$$= \frac{1}{q\Gamma(\pi^\ell)} \sum_{j=1}^{q-1} \pi^\ell(\epsilon^j) f(x-\epsilon_k^j, k)$$

$$= \frac{1}{q\Gamma(\pi^\ell)} \sum_{j=1}^{q-1} \pi^\ell(\epsilon^j) d_k f(x-\epsilon_k^j) \ ,$$

since $f(x-\epsilon_k^j, k+1)$ is contant over $j = 1,2,\dots,q-1$.

Also,

$$(2.2') \qquad T_o d_k f(x) = d_k f(x).$$

Fix a coset $y + \mathfrak{P}^{-(k+1)}$ and make the conventions, $\epsilon^o = \epsilon_k^o = 0$ and $\pi(0) = \pi^\ell(0) = 0$. Let,

$$(2.3) \qquad \begin{cases} a_o = f(y,k+1) & \alpha_o^j = d_k f(y+\epsilon_k^j) \\ a_\ell = T_\ell f(y,k+1) & \alpha_\ell^j = T_\ell d_k f(y+\epsilon_k^j) \end{cases}$$

$\ell = 1,2,\dots,q-2;\ j = 0,1,\dots,q-1.$

Then for $x \in y + \epsilon_k^j + \mathfrak{P}^{-k}$, (2.2) can be rewritten, for $\ell = 1,\dots,q-2$,

$$(2.4) \qquad \begin{cases} \alpha_\ell^j = \dfrac{1}{q\Gamma(\pi^\ell)} \displaystyle\sum_{k=0}^{q-1} \pi^\ell(\epsilon^j - \epsilon^i) d_k f(y+\epsilon_k^i) \\ \quad = \dfrac{1}{q\Gamma(\pi^\ell)} \displaystyle\sum_{i=0}^{q-1} \pi^\ell(\epsilon^j - \epsilon^i)\alpha_0^i \end{cases}$$

Namely, if $\{\alpha_0^i\}$ are given then $\{\alpha_\ell^i\}$ are obtained by a linear transformation whose matrix has (i,j) entry $(1/q\,\Gamma(\pi^\ell))\pi^\ell(\varepsilon^i-\varepsilon^j)$. Since regularity and sub-regularity are defined in terms of the local behaviour of the $\{\alpha_\ell^j\}$ and $\{\alpha_0^j\}$ we see that we are headed towards a local definition of our conjugacy relations.

Example: K, the 3-adic or 3-series number field.

We have $q = 3$, $q-1 = 2$ so $\pi(\varepsilon)$ is a primitive 2nd root of unity; namely, $\pi(\varepsilon) = -1$. ε can be chosen to be -1. Thus, we have (with our convention above), $\pi(0) = 0$, $\pi(\varepsilon) = -1$, $\pi(\varepsilon^2) = 1$. We evaluate $\Gamma(\pi)$.

$$\Gamma(\pi) = \int_{|x|=3} \pi(x)\,\bar{\chi}(x)|x|^{-1}dx = (1/3)\int_{|x|=3} \pi(x)\,\bar{\chi}(x)dx$$

$$= (1/3)(\bar{\chi}(p^{-1}) - \bar{\chi}(-p^{-1})) = (1/3)\overline{(\chi(p^{-1}) - \chi(-p^{-1}))}$$

$\chi(p^{-1})$ is a primitive 3rd root of unity and $\chi(-p^{-1})$ is its square. Thus $\Gamma(\pi) = (1/3)(\pm\, i\sqrt{3}\,)$, and so $(1/q\,\Gamma(\pi)) = \pm\,(1/i\sqrt{3}\,)$. The choice is not essential, let us take $1/3\,\Gamma(\pi) = 1/i\sqrt{3}$.

If now $\alpha_o = (\alpha_o^1, \alpha_o^o, \alpha_o^{-1})$ (we require only that $\alpha_o^1 + \alpha_o^o + \alpha_o^{-1} = 0$) then $\alpha_1 = (\alpha_1^1, \alpha_1^o, \alpha_1^{-1})$ is given by $\alpha_1^o = (1/i\sqrt{3}\,)(\alpha_o^{-1} - \alpha_o^1)$, $\alpha_1^1 = (1/i\sqrt{3}\,)(\alpha_o^o - \alpha_o^{-1})$, $\alpha_1^{-1} = (1/i\sqrt{3}\,)(\alpha_o^1 - \alpha_o^o)$.

Notation. For $c = (c_0, c_1, \ldots, c_{q-1}) \in C^q$, let $\|c\| = \left[\sum_{s=0}^{q-1} |c_s|^2\right]^{1/2}$.

Proposition (2.5). *Let* $(\alpha_o^o, \alpha_o^1, \ldots, \alpha_o^{q-1}) \in C^q$ *be given and let* $\alpha_\ell = (\alpha_\ell^o, \alpha_\ell^1, \ldots, \alpha_\ell^{q-1}) \in C^q$, $\ell = 1, \ldots, q-2$ *be defined as in* (2.3) *and* (2.4). *Then,*

(a) $\sum_{j=0}^{q-1} \alpha_\ell^j = 0$, $\ell = 0, 1, \ldots, q-2$

(b) $\|\alpha_\ell\| = \|\alpha_o\|$, $\ell = 1, 2, \ldots, q-2$

(c) $\sum_{i=0}^{q-1} \alpha_j^i \, \alpha_\ell^i = 0$, *whenever* $j + \ell$ *is odd*;

$j, \ell = 0, 1, \ldots, q-2$.

Proof. (a) This either follows from regularity or we may start with $\sum_{i=0}^{q-1} \alpha_o^i = 0$ from regularity, or by assumption. We then let α_ℓ be defined by (2.4), $\ell \neq 0$, and the rest of the proof will follow. Thus, if $\ell \neq 0$

$$\sum_{j=0}^{q-1} \alpha_\ell^j = (1/q\,\Gamma(\pi^\ell)) \sum_{j=0}^{q-1} \sum_{i=0}^{q-1} \pi^\ell(\epsilon^j - \epsilon^i)\alpha_o^i$$

$$= (1/q\,\Gamma(\pi^\ell)) \sum_{i=0}^{q-1} \alpha_o^i \Big(\sum_{j=0}^{q-1} \pi^\ell(\epsilon^j - \epsilon^i) \Big)$$

$$= 0, \quad \text{since} \sum_{j=0}^{q-1} \pi^\ell(\epsilon^j - \epsilon^i) = 0, \text{ all } i.$$

Note that π^{ℓ} is a non-trivial character on $\mathfrak{O}^*$ and is ramified of degree 1 and $\{\epsilon^j-\epsilon^i\}_{j=0}^{q-1}$ are a complete set of coset representatives of $\mathfrak{P}$ in $\mathfrak{O}$, recalling our convention that $\epsilon^o = 0$!! This proves (a).

(b) Let g be the restriction of $d_k f$ to $y + \mathfrak{P}^{-(k+1)}$. Using (2.2) and (2.1) we see that $T_{\ell} g = T_{\ell} d_k f$ on $y + \mathfrak{P}^{-(k+1)}$, and it is supported on $y + \mathfrak{P}^{-(k+1)}$.

Note that $\int g = q^k \sum_{j=0}^{q-1} \alpha_o^j = 0$, and $\left[\int |g|^2\right]^{\frac{1}{2}} = q^{k/2}\left[\sum_{j=0}^{q-1} |\alpha_o^j|^2\right]^{\frac{1}{2}}$ $= q^{k/2}\|\alpha_o\|$. Similarly, $\left[\int |T_{\ell} g|^2\right]^{\frac{1}{2}} = q^{k/2}\|\alpha_{\ell}\|$. Thus,

$$q^{k/2}\|\alpha_{\ell}\| = \left[\int |T_{\ell} g|^2\right]^{\frac{1}{2}} = \left[\int |(T_{\ell} g)^\wedge|^2\right]^{\frac{1}{2}} = \left[\int |\pi^{-\ell}\hat{g}|^2\right]^{\frac{1}{2}} = \left[\int |\hat{g}|^2\right]^{\frac{1}{2}}$$

$$= \left[\int |g|^2\right]^{\frac{1}{2}} = q^{k/2}\|\alpha_o\|.$$ This proves (b).

(c) We use g above, but for simplicity we assume that $y = 0$. (This amounts to multiplying $\hat{g}$ by $\bar{\chi}_y$.)

g is supported on $\mathfrak{P}^{-(k+1)}$ and is constant on cosets of $\mathfrak{P}^{-k}$, so $\hat{g}$ is supported on $\mathfrak{P}^k$ and is constant on cosets of $\mathfrak{P}^{k+1}$. Since $\hat{g}(0) = 0$, we see that $\hat{g}$ is constant on each of the cosets $\left\{\epsilon^j_{-(k+1)} + \mathfrak{P}^{k+1}\right\}_{j=1}^{q-1}$ and is zero otherwise. $(T_{\ell} g)^\wedge(\epsilon^j_{-(k+1)})$ $= \pi^{-\ell}(\epsilon^j)\hat{g}(\epsilon^j_{-(k+1)})$. Thus,

$$q^k \sum_{i=0}^{q-1} \alpha_j^i \alpha_\ell^i = q^k \sum_{i=0}^{q-1} T_j g(\epsilon_k^i) T_\ell g(\epsilon_k^i)$$

$$= \int_K T_j g(x) T_\ell g(x)\,dx$$

$$= \int_K (T_j g)\hat{}(\xi)(T_\ell g)\hat{}(-\xi)\,d\xi$$

$$= q^{-(k+1)} \sum_{i=1}^{q-1} \pi^{-j}(\epsilon^i) g(\epsilon^i_{-(k+j)}) \pi^{-\ell}(-\epsilon^i) \hat{g}(-\epsilon^i_{-(k+1)})$$

$$= q^{-(k+1)} \pi^\ell(-1) \sum_{i=j}^{q-1} \pi^{-(j+\ell)}(\epsilon^i) \hat{g}(\epsilon^i_{-(k+1)}) \hat{g}(-\epsilon^i_{-(k+1)}) .$$

Since $(j+\ell)$ is odd, $\pi^{-(j+\ell)}(-\epsilon^i) = -\pi^{-(j+\ell)}(\epsilon^i)$, and it follows easily that the sum is zero.

The next result shows the importance of the three properties exhibited in (2.5).

<u>Theorem (2.6)</u>. <u>Consider a</u> $k \times (m+1)$ <u>matrix</u> (α_j^i) <u>with complex entries;</u> $i = 0,\ldots,k-1$; $j = 0,\ldots,m$.

<u>Let</u> $\alpha_j = (\alpha_j^o, \alpha_j^1, \ldots, \alpha_j^{k-1}) \in C^k$, $\|\alpha_j\| = \left[\sum_{i=0}^{k-1} |\alpha_j^i|^2\right]^{1/2}$

$$\alpha^i = (\alpha_o^i, \alpha_1^i, \ldots, \alpha_m^i) \in C^{m+1}, \ \|\alpha^i\| = \left[\sum_{j=0}^{m} |\alpha_j^i|^2\right]^{1/2}$$

$$a = (a_o, a_1, \ldots, a_n) \in C^{m+1}, \ \|a\| = \left[\sum_{i=0}^{m} |a_j|^2\right]^{1/2} .$$

Suppose $\{0,1,\ldots,m\} = D \cup E$, *where* D *and* E *are non-empty and disjoint. Suppose now that,*

$$(2.7)\qquad \sum_{i=0}^{k-1} \alpha_j^i = 0\ ,\quad j = 0,1,\ldots,m$$

$$(2.8)\qquad \|\alpha_j\| = \|\alpha_o\|,\quad j = 1,2,\ldots,m$$

$$(2.9)\qquad \sum_{i=0}^{k-1} \alpha_j^i\, \alpha_\ell^i = 0 \quad \text{whenever} \quad j \in D,\ \ell \in W.$$

Then there is a p_o, $0 < p_o < 1$ *such that*

$$(2.10)\qquad \|a\|^p \le (1/k) \sum_{i=0}^{k-1} \|a + \alpha^i\|^p$$

for all $p \ge p_o$, *where* p_o *depends only on* k *and* m.

The proof is delayed. It starts with the proof of Lemma (2.12).

Definition. Let $F(x,k) = (f_o(x,k), f_1(x,k), \ldots, f_m(x,k))$ be a vector of smooth functions defined on a domain $\mathcal{D} \subset K^n \times \mathbb{Z}$. For each $(y,k+1) \in \mathcal{D} \sim \partial.\mathcal{D}$ let

$$\left.\begin{aligned} a_j &= f_j(y,k+1) \\ a_j^i &= d_k f_j(y+e_k^i) \end{aligned}\right\}\qquad \begin{aligned} j &= 0,1,\ldots,m \\ i &= 0,\ldots\ ,q^n-1 \end{aligned} \quad .$$

If the (α_j^i) satisfy (2.7), (2.8) and (2.9) we say that $F(x,k)$ is a *conjugate system* on $\mathcal{D}$.

Remark. (2.7) simply says that each $f_j(x,k)$ is regular on $\mathcal{D}$, and the conclusion of (2.6); namely (2.10), say that $|F(x,k)|^p$ is sub-regular. Thus,

Corollary (2.11). *If* $F = (f_0, f_1, \ldots, f_m)$ *is a conjugate system on a domain* $\mathcal{D}$, *then there is a* p_0, $0 < p_0 < 1$, p_0 *independent of* F, *such that* $|F(x,k)|^p$ *is sub-regular on* $\mathcal{D}$ *for all* $p \geq p_0$.

Examples of Conjugate Systems.

Using (2.4) we see that if $d_k f$ is defined on a domain, then $d_k T_\ell f$ is defined on that same domain, without reference to the global existence of $T_\ell f$. Since $d_k f$ is just the derived function of f (see IV §3), if the domain $\mathcal{D}$ is connected then f is determined up to a constant by $d_k f$. Thus, if f is defined on a connected domain $\mathcal{D}$ then, taking this point of view, $T_\ell f$ is defined on $\mathcal{D}$, and is uniquely determined, up to an additive constant.

Thus, let $\{\ell_j\}_{j=1}^m$ be a subset of $\{0,1,\ldots,q-2\}$, with at least one odd and one even integer. If D is the odd integers and E the even integers in $\{\ell_j\}$ then by (2.5) and (2.6), $(T_{\ell_1} f, T_{\ell_2} f, \ldots, T_{\ell_m} f)$ is a conjugate system on $K \times \mathbb{Z}$, or any connected domain in $K \times \mathbb{Z}$.

In particular, (f,T_1f), $(f,T_1f,\ldots,T_{q-2}f)$ are called the <u>principal</u> and <u>full</u> conjugate systems, respectively.

Recall that q is odd. Then $\pi^{(q-1)/2}(\epsilon^s) = \zeta^{((q-1)/2)\cdot s} = (-1)^s$. If $(q-1)/2$ is odd, then $(f,T_{(q-1)/2}f)$ is a conjugate system, $T_{(q-1)/2}f$ is a natural analogue of the Hilbert transform of f.

<u>Lemma (2.12).</u> <u>Let</u> a <u>and</u> (α_j^i) <u>be given as in (2.6) so that (2.7), (2.8) and (2.9) are valid. Then given</u> $p_1 > (m-1)/m$, <u>there is a constant</u> $A_{p_1} > 0$ <u>such that</u> (2.10) <u>holds for all</u> $p > p_1$ <u>provided</u> $\|\alpha_o\| \leq A_{p_1}\|a\|$.

<u>Proof.</u> We may assume that $\|a\| \neq 0$, and $0 < p_1 \leq p \leq 1$, for if $\|a\| = 0$, $p_1 > 1$ or $p > 1$ the result is trivial.

$$(2.13)\quad \left\{\begin{aligned} \sum_{i=0}^{k-1}\|a+\alpha^i\|^p &= \sum_{i=0}^{k-1}\Big\{\sum_{j=0}^{m}|a_j+\alpha_j^i|^2\Big\}^{p/2} \\ &= \sum_{i=0}^{k-1}\Big\{\sum_{j=0}^{m}|a_j|^2 + 2\,\mathrm{Re}\sum_{j=0}^{m}\bar{a}_j\alpha_m^i + \sum_{j=0}^{m}|\alpha_j^i|^2\Big\}^{p/2} \\ &= \sum_{i=0}^{k-1}\Big\{\|a\|^2 + 2\,\mathrm{Re}\sum_{j=0}^{m}\bar{a}_j\alpha_j^i + \|\alpha^i\|^2\Big\}^{p/2} \\ &= \|a\|^p\sum_{i=0}^{k-1}\Big\{1 + \frac{2\,\mathrm{Re}\sum_{j=0}^{m}\bar{a}_j\,\alpha_j^i}{\|a\|^2} + \frac{\|\alpha^i\|^2}{\|a\|^2}\Big\}^{p/2} \end{aligned}\right.$$

Using (2.8) and (2.9) we obtain,

$$(2.14)\quad \begin{cases} \sum_i |\mathrm{Re}\sum_j \bar{a}_j \alpha_j^i|^2 \le \sum_i |\sum_{j\in D} \bar{a}_j\alpha_j^i + \sum_{j\in E} a_j \overline{\alpha_j^i}|^2 \\ \quad = \sum_i |\sum_{j\in D} \bar{a}_j\alpha_j^i|^2 + \sum_i |\sum_{j\in E} a_j\overline{\alpha_j^i}|^2 \\ \qquad + 2\,\mathrm{Re}\sum_{j\in D;\,k\in E} \overline{a_j a_k}\left(\sum_i \alpha_j^i\alpha_k^i\right) \\ \quad \le \sum_i\left(\sum_{j\in D}|a_j||\alpha_j^i|\right)^2 + \sum_i\left(\sum_{j\in E}|a_j||\alpha_j^i|\right)^2 \\ \quad \le \sum_i\left(\sum_{j\in D}|a_j|^2\right)\left(\sum_{j\in D}|\alpha_j^i|^2\right) \\ \qquad + \sum_i\left(\sum_{j\in D}|a_j|^2\right)\left(\sum_{j\in E}|\alpha_j^i|^2\right) \\ \quad \le m\|\alpha_o\|^2\left(\sum_{j\in D}|a_j|^2\right) + m\|\alpha_o\|^2\left(\sum_{j\in E}|a_j|^2\right) \\ \quad = m\|\alpha_o\|^2\,\|a\|^2 . \end{cases}$$

In particular, $|\mathrm{Re}\sum_j \bar{a}_j\ \alpha_j^i| \le \sqrt{m}\ \|\alpha_o\|\cdot\|a\|$. Note that $\|\alpha^i\|^2 \le (m+1)\|\alpha_o\|^2$. Suppose $\|\alpha_o\| \le \|a\|/3\sqrt{m}$.

Then, $|(2\ \mathrm{Re}\sum_j \bar{a}_j\alpha_j^i)/\|a\|^2 + \|\alpha^i\|^2/\|a\|^2|$

$$\le (2\sqrt{m}\ \|\alpha_o\|\cdot\|a\|)/\|a\|^2 + (m+1)\|\alpha_o\|^2)/\|a\|^2$$

$$\leq \; 2/3 + \frac{m+1}{9m}$$

$$\leq \; 8/9 < 1.$$

Hence, we may use the binomial theorem on each summand in (2.13) and obtain:

$$(2.15)\quad \left\{ \begin{aligned} &\sum_i \|a+\alpha^i\|^p \\ &\quad = \|a\|^p \sum \Big\{1 + \frac{p \ \mathrm{Re} \sum_j \bar{a}_j \alpha_j^i}{\|a\|^2} + \frac{p}{2} \cdot \frac{\|\alpha^i\|^2}{\|a\|^2} \\ &\quad - \frac{p(2-p)}{8\|a\|^4} \Big[4(\mathrm{Re} \sum_j \bar{a}_j \alpha_j^i)^2 + \|\alpha^i\|^4 \\ &\quad + 4\|\alpha^i\|^2 (\mathrm{Re} \sum_j \bar{a}_j \alpha_j^i) \Big] + R_{3i} \Big\}, \end{aligned} \right.$$

where R_{3i} are the third Taylor remainders.

Observe that,

$$\sum_i \frac{p \ \mathrm{Re} \sum_j a_j \alpha_j^i}{\|a\|^2} = (p/\|a\|^2) \ \mathrm{Re} \sum_j \bar{a}_j \Big(\sum_i \alpha_j^i \Big) = 0 \qquad (2.7)$$

$$p/2 \sum_i \|\alpha^i\|^2/\|a\|^2 = ((m+1)p/2)\|\alpha_o\|^2/\|a\|^2 \qquad (2.8)$$

$$\begin{aligned} &(p(2-p)/8\|a\|^4) \sum_i 4(\mathrm{Re}(\sum \bar{a}_j \alpha_j^i))^2 \\ &\qquad \leq (p(2-p)/2\|a\|^4)(m\|\alpha_o\|^2 \cdot \|a\|^2 \\ &\qquad \leq (mp(2-p)/2)\|\alpha_o\|^2/\|a\|^2 \qquad (2.14), \end{aligned}$$

and that the remaining terms are bounded by $B(\|\alpha_o\|/\|a\|)^3$, where $B > 0$ is independent of p.

If we enter these results in (2.15) we get

$$\sum_{i=0}^{k-1} \|a+\alpha^i\|^p \geq \|a\|^p\left\{k + \frac{m+1}{2}\, p\, \frac{\|\alpha_o\|^2}{\|a\|^2} - m\,\frac{p(2-p)}{2}\,\frac{\|\alpha_o\|^2}{\|a\|^2} - B\left(\frac{\|\alpha_o\|}{\|a\|}\right)^3\right\}$$

$$= \|a\|^p\left\{k + \left(\frac{\|\alpha_o\|}{\|a\|}\right)^2\left[\frac{mp}{2}\left(p - \frac{(m-1)}{m}\right) - B\,\frac{\|\alpha_o\|}{\|a\|}\right]\right\}.$$

Thus, $(1/k)\sum_{i=0}^{k-1} \|a+\alpha^i\|^p \geq \|a\|^p$, provided,

$p \geq p_1 > (m-1)/m$ and $(\|\alpha_o\|/\|a\|) \leq A_{p_1}$, where

$$A_{p_1} = \min\{1/3\sqrt{m}, p_1(mp_1-(m-1))/(2B)\}.$$

This completes the proof of (2.12).

<u>Proof of (2.6)</u>. If $\|a\| = 0$ the result holds trivially so we may assume that $\sum_i \|a+\alpha^i\| > 0$ (since, $k\|a\| \leq \sum_i \|a+\alpha^i\|$). Also, in view of (2.12) we may fix p_1, $(m-1)/m < p_1 < 1$ and assume that $\|\alpha_o\| \geq A_{p_1}\|a\|$.

Thus we may assume,

$$(2.16) \qquad (1/k)\sum_i \|a+\alpha^i\| = 1,\ \|\alpha_o\| \geq A_{p_1}\|a\|,\ A_{p_1} > 0 .$$

Note that $\|\alpha_o\| = 0$ is not consistent with (2.16). For if $\|\alpha_o\| = 0$ then $\|\alpha_j\| = 0$ for all j and $\alpha_j^i = 0$ for all i and j, and so $\alpha^i = 0$ all i and thus $\|a\| = (1/k)\sum_i \|a+\alpha^i\| = 1$, and $0 = \|\alpha_o\| \geq A_{p_1}\|a\|$. But $A_{p_1} > 0$ so $\|a\| = 0$, a contradiction.

Let B be the collection of all vectors $b = \{a+\alpha^i\}_{i=0}^{k-1}$ satisfying (2.7), (2.8), (2.9) and (2.16). B is a compact set in $C^{k(m+1)}$.

To prove the theorem it will be sufficient to show that there is a δ, $0 < \delta < 1$ such that

$$\text{(2.17)} \qquad \|a\| \leq \delta(1/k)\sum_{i=0}^{k-1} \|a+\alpha^i\|, \quad \text{all } \{a+\alpha^i\} \in B.$$

For if (2.17) holds, then (2.10) is valid for all $p \geq p_2 = (1 + \log_k 1/\delta)^{-1}$ and $\{a+\alpha^i\} \in B$, and (2.10) is then valid for all $\{a+\alpha^i\}$ and $p > p_o = \max[p_1, p_2]$, and we see that $p_o < 1$.

To prove (2.17) for $\{a+\alpha^i\} \in B$ we note that if (2.17) fails, then since B is compact there is $\{a+\alpha^i\} \in B$ such that

$$\|a\| = (1/k)\sum_{i=0}^{k-1} \|a+\alpha^i\|.$$

We see that in this case there are real numbers λ_i, $i = 0,1,\ldots,k-1$ such that $a + \alpha^i = \lambda_i\alpha$, $i = 0,\ldots,k-1$. (The

possibility that $a \equiv 0$ can be excluded since it leads to an obvious contradiction.) For all i and j we then have
$\alpha_j^i = (\lambda_i - 1)a_j$.

Now take $j \in D$ and $k \in E$. We see that

$$0 = \sum_i \alpha_j^i \alpha_k^i = a_j a_k \sum_i (\lambda_i - 1)^2 \qquad (2.9).$$

According to this, either $a_j = 0$, $a_k = 0$ or $\lambda_i = 1$ all i. If $a_j = 0$ then $\alpha_j^i = 0$ all i, so $\|\alpha_j\| = 0$ and $\|\alpha_o\| = 0$. Similarly $a_k = 0$ implies $\|\alpha_o\| = 0$. If $\lambda_i = 1$ all i then $\alpha_i^j = 0$ all i and j, and again $\|\alpha_o\| = 0$. But we noted above that $\|\alpha_o\| = 0$ leads to a contradiction.

3. F. and M. Riesz theorem

Theorem (3.1). *Suppose* $F(x,k) = (f_o(x,k),\ldots,f_m(x,k))$ *is a congugate system on* $K^n \times \mathbb{Z}$ *and* $\int_{K^n} |F(x,k)|\,dx \leq A$ *for all* $k \in \mathbb{Z}$, $A > 0$ *independent of* k. *Then* $\lim_{k \to -\infty} F(x,k) = F(x)$ *exists in* L^1 *and a.e. and* $f_j^*(x) = \sup_k |f_j(x,k)| \in L^1$, $j = 0,1,\ldots,m$.

Proof. Using (2.11) we see that $|F(x,k)|^p$ is sub-regular for some $p < 1$. The theorem follows from an application of (1.1).

Corollary (3.2). *Suppose* $\mu_0, \mu_1, \ldots, \mu_m$ *are finite Borel measures on* $K^n \times \mathbb{Z}$. *If* $F(x,k) = (\mu_0(x,k), \ldots, \mu_m(x,k))$ *is a conjugate system then each* μ_ℓ *is absolutely continuous.*

Proof. $$\int_{K^n} |F(x,k)|\,dx = \int_{K^n} \Big(\sum_j |\mu_j(x,k)|^2\Big)^{1/2} dx$$

$$\leq \int_{K^n} \Big(\sum_{j=0}^{m} |\mu_j(x,k)|\Big) dx$$

$$= \sum_{j=0}^{m} \|\mu_j(\cdot,k)\|_1 \leq \sum_{j=0}^{m} \|\mu_j\|_M .$$

We apply (3.1) and from IV (1.9)(c) we see that $\mu_j(x,k)$ is the regularization of $f_j \in L^1$.

Theorem (3.3). (F. and M. Riesz theorem). *Suppose* μ *is a finite Borel measure and* π *is a unitary multiplicative character on* K^* *which is odd, homogeneous of degree zero and ramified of degree 1. Suppose* $\hat{\mu}$ *is supported on a set* $\{x : \pi(x) = c\}$, *where* c *is some constant. Then* μ *is absolutely continuous.*

Proof. Let $\tilde{f}(x,k)$ be defined by $(\tilde{f}(\cdot,k))^\wedge = \hat{\mu}\,\Phi_k \pi^{-1}$. Then $(f,\tilde{f})$ is a conjugate system where $f(x,k) = \mu(x,k)$. But $\tilde{f}(x,k) = (c^{-1}\hat{\mu}\,\Phi_k)^\vee = c^{-1} f(x,k)$ and so $|F(x,k)| = |(f(x,k), \tilde{f}(x,k))|$ $= \sqrt{2}\,|f(x,k)|$, $\int |F(x,k)|\,dx = \sqrt{2}\int |f(x,k)| \leq \sqrt{2}\,\|\mu\|_M$. Apply (3.2).

4. Notes for Chapter VII

General references for this chapter are Chao and Taibleson [1] and Chao [1], [2].

In this chapter we are concerned with developing the notion of conjugate systems on local fields; describing their properties and exhibiting examples. Our approach borrows from the ideas used by Coifman and Weiss [1] in their study of generalized Cauchy-Riemann systems. Our systems are described locally by requiring that the system satisfy certain linear difference equations which are analogues of the Cauchy-Riemann equations.

The principal conjugate system described (following (2.11)) requi that the module of the field K, q, be odd, and that π be an odd character, ramified of degree 1. These seem to be essential requireme in the following sense: In some recent work (unpublished) we have constructed examples of finite Borel measures, μ, such that $T_{\pi}\mu = \mu$, and such that μ is not absolutely continuous. We have examples for the case $q = 4$, π ramified of degree 1 and primitive; and for $q = 5$, π ramified of degree 1 and even.

An important consequence of the existence of the conjugate system described following (2.11) is the following: Let $F = (f_0, f_1, \ldots, f_m)$

be such a conjugate system and let p_0, $0 < p_0 < 1$ be given as in that statement. We say that $f \in \mathcal{S}'$ is in $H^p(K^n)$, $p > 0$ iff $\int (\sup_k |f(x,k)|)^p dx < \infty$. If $p > p_0$, then $f \in H^p$ iff f is a component of a conjugate system F, where $\sup_k \int |F(x,k)|^p dx < \infty$. The "if" part follows from (2.11) and (1.1)(c), the "only if" part follows from a result of Chao [2] that for π any multiplicative character on $\mathfrak{O}^*$, then $f \in H^p$ implies $T_\pi f \in H^p$.

Chapter VIII. Almost everywhere convergence of Fourier series

1. The results

Theorem (1.1). *If* $f \in L^p(\mathfrak{D})$, $1 < p < \infty$, *and* $\mathfrak{M} f(x) = \sup_n |S_n f(x)|$, *then* $\mathfrak{M} f \in L^p$ *and there is a constant* $A_p > 0$ *independent of* f *such that* $\|\mathfrak{M} f\|_p \leq A_p \|f\|_p$.

Corollary (1.2). *If* $f \in L^p(\mathfrak{D})$, $1 < p \leq \infty$, *then* $S_n f(x) \to f(x)$ a.e. *as* $n \to \infty$

Proof. We may assume $1 < p < \infty$. Since $\mathcal{S}$ is dense in L^p, we approximate f in $\mathcal{S}$ and then use (1.1). Thus $f = b+g$, $\|b\|_p < \epsilon$, $g \in \mathcal{S}$. $S_n g = g$ for n large enough, and thus for n large enough, $\limsup |S_n f(x) - f(x)| \leq \limsup |S_n b(x)| + |b(x)| \leq \mathfrak{M} b(x) + |b(x)|$. By an obvious estimate $|\{x : \limsup |S_n f(x) - f(x)| > \delta\}|$ $\leq ((1 + A_p)\, \epsilon\, \delta^{-1})^p$, etc.

Basic Result (1.3). *Let* $F \subset \mathfrak{D}$ *be a measurable subset of* $\mathfrak{D}$. *Let* f *be the characteristic function of* F. *There is a constant* $C_p > 0$ *independent of* F *and* $y > 0$, $C_p \leq C\, p^2/(p-1)$ (C *depends only on* K) *such that*

$$|\{x \in \mathfrak{D} : \mathfrak{M} f(x) > y\}| \leq C_p^p\, y^{-p} |F| .$$

If $p = 2$, f *may be any function in* L^2.

Remarks. The entirety of §3 is concerned with the proof of (1.3). (1.1) is a consequence of (1.3) by purely measure theoretic considerations. What is required is the concept of restricted weak type and an appropriate extension of the Marcinkiewicz interpolation theorem. These points are discussed in the Notes on this chapter in §4.

2. Notation and preliminary results

We review and extend a variety of results about Fourier series. $\{\chi_n\}$ is the collection of characters on Ω.

(2.1) If $n = \ell q^\nu + s$, $0 \leq s \leq q^\nu$, ℓ a non-negative integer then $\chi_n = \chi_{\ell q^\nu} \cdot \chi_s$

(2.2) If $0 \leq s < q^\nu$ then χ_s is constant on spheres of radius less than or equal to $q^{-\nu}$, $\nu = 0,1,\ldots$.

Definition. If ω is a sphere in Ω, n a non-negative integer we let $n[\omega]$ be the largest integer in $n|\omega|$.

Notice that $n = n[\omega] \cdot |\omega|^{-1} + s$, $0 \leq s < |\omega|^{-1}$.

In view of (2.1) and (2.2),

(2.3) $$\left| \frac{1}{|\omega|} \int_{\omega} \chi_n(t)\overline{\chi}_m(t)dt \right| = \begin{cases} 1, & n[\omega] = m[\omega] \\ 0, & n[\omega] \neq m[\omega] \end{cases}$$

and

(2.4) $\{|\omega|^{-1/2}\chi_{n|\omega|^{-1}}\}_{n=0}^{\infty}$ _is a complete orthonormal system on_ ω.

_If_ $0 \in \omega$, _it is a complete set of characters on_ ω, ω _being an additive group_.

If $f \in L^1(\omega)$, $\omega \subset \mathfrak{O}$ we define Fourier coefficients of f on ω by

$$c_n(\omega;f) = \frac{1}{|\omega|} \int_{\omega} f(t)\overline{\chi}_{n|\omega|^{-1}}(t)dt, \quad n = 0,1,\ldots .$$

Plancherel's formula gives

(2.5) $$\sum_{n=0}^{\infty} |c_n(\omega;f)|^2 \, |\omega| = \int_{\omega} |f(x)|^2 dx .$$

For $\omega \subset \mathfrak{O}$, $\omega = x + \mathfrak{P}^k$, define $\omega^* = x + \mathfrak{P}^{k-1}$. Thus, $\mathfrak{O}^* = \mathfrak{P}^{-1}$ (and for this chapter $\mathfrak{O}^* \neq \{|x| = 1\}$, but $\mathfrak{O}^* = \{|x| \leq q\}$). We extend f to $\mathfrak{P}^{-1}$ by setting $f(x) = 0$ if $|x| = q$ and similarly we extend characters on $\mathfrak{O}$ to $\mathfrak{P}^{-1}$.

For all $\omega^* \subset \mathfrak{O}^*$ there are exactly q spheres $\overline{\omega}$ such that $(\overline{\omega})^* = \omega^*$.

For each $\omega \subset \mathfrak{O}$ there is a unique sequence $\{\omega_j\}_{j=0}^{J}$, with

$\omega_0 = \mathfrak{O}$, $\omega_J = \omega$ and $\omega_j^* = \omega_{j-1}$, $j = 1,2,\ldots J$. For this sequence, $|\omega_j| = q^{-j}$, $|\omega_j^*| = q^{-j+1}$.

Recall the definition of $D_n^{\#}$ (VI, §3) and the convention that $D_0 = D_0^{\#} \equiv 0$. $D_n^{\#} = \bar{\chi}_n D_n$, $D_n = \sum_0^{n-1} \chi_\nu$.

<u>Definition.</u> For $\omega^* \subset \mathfrak{P}^{-1}$, $n \geq 0$, $x \in \omega^*$

$$S_n^{\#}f(x;\omega^*) = \int_{\omega^*} f(t)\bar{\chi}_n(t)D_n^{\#}(x-t)dt.$$

(2.6) $$|S_n^{\#}f(x;\omega^*)| = |\int_{\omega^*} f(t)D_n(x-t)dt|.$$

Obvious and trivial.

(2.7) <u>If</u> ω <u>is any sphere in</u> $\mathfrak{P}^{-1}$, <u>and</u> $x \notin \omega$ <u>then</u> $D_n^{\#}(x-t)$ <u>is constant as</u> t <u>varies over</u> ω.

<u>Proof</u>. This reduces to $D_n^{\#}(x+y) = D_n^{\#}(x)$ if $|y| < |x|$. That is the conclusion of VI (3.3) for $D_n^{\#}$ on $\mathfrak{O}$, and clearly its extension to $\mathfrak{P}^{-1}$ has the same properties.

<u>Definition.</u> $C_n(\omega^*;f) = \max[|c_n(\bar{\omega}^*;f)| : \bar{\omega}^* = \omega^*]$.

(2.8) <u>If</u> $n[\omega] = m[\omega]$ <u>and</u> $x \in \mathfrak{P}^{-1}$ <u>then</u>

$$|S_n^{\#}f(x;\omega^*)| \leq |S_m^{\#}f(x;\omega^*)| + q\, C_{n[\omega]}(\omega^*;f).$$

Proof. $\left| |S_n f(x;\omega^*)| - |S_m f(x;\omega^*)| \right|$

$$\leq \left|\int_{\omega^*} f(t)(D_n(x-t) - D_m(x-t))dt\right|$$

$$\leq \sum_{\bar{\omega} \ni \bar{\omega}^* = \omega^*} \left|\int_{\bar{\omega}} f(t)(D_n(x-t) - D_m(x-t))dt\right|$$

$$\leq \sum_{\substack{\bar{\omega} \ni \\ \bar{\omega}^* = \omega^*}} \sum_{s=0}^{|\omega|^{-1}-2} \left|\int_{\bar{\omega}} f(t)\chi_{n[\omega]|\omega|^{-1}+s}(x-t)dt\right|$$

$$= \sum_{\bar{\omega} \ni \bar{\omega}^* = \omega^*} \sum_{s=0}^{|\omega|^{-1}-2} C_{n[\omega]}(\bar{\omega};f)$$

$$\leq q(|\omega|^{-1}-1)|\omega|\ C_{n[\omega]}(\omega^*;f) < q\ C_{n[\omega]}(\omega^*;f).$$

Remark. On occasion ω^* is given explicitly while ω is not so given, but we still wish to write $n[\omega]$. In such a case $n[\omega] \equiv n[\bar{\omega}]$ where $(\bar{\omega})^* = \omega^*$, or what amounts to the same thing, $n[\omega]$ is the largest integer in $n|\omega^*|^{-1} q$.

(2.9) Suppose $g \in L^\infty(\omega^*)$ and $\|g\|_{\infty;\omega^*}$ is the ess sup of g on ω^*. For $x \in \omega^* \subset \mathfrak{P}^{-1}$ define

$$T_n^\# g(x;\omega^*) = \sup_{x \in \tilde{\omega}^* \subset \omega^*} \left|\int_{\omega^* \sim \tilde{\omega}^*} g(t) D_n^\#(x-t)dt\right| .$$

If $y > 0$ then there is an $A > 0$, independent of g, ω^* and y such that

$$|\{x \in \omega^*: T_n^{\#}g(x;\omega^*) > y\}| \leq e^2[\exp\{-Ay/\|g\|_{\infty;\omega^*}\}]|\omega^*| .$$

Proof. This is VI (3.7).

3. Proof of the Basic Result

A. Reduction to the Basic Lemma

Basic Lemma (3.1). *Fix* $y > 0$, $1 < p < \infty$, N *a positive integer and* F *a measurable subset of* $\mathfrak{D}$ *with* f *its characteristic function. Let*

$$(3.2) \qquad L = L(p) = [2\,p^2/(p-1)] + 1.$$

For $p = 2$ *we let* $L(2) = 8$ *and* f *can be any function in* L^2.

Then there is a set $E = E(y,p,N,f) \subset \mathfrak{D}$ *and a constant* $C > 0$ *that depends on* K *but is independent of* y, p, N *and* f *such that* $|E| \leq C^p y^{-p} \int_{\mathfrak{D}} |f(x)|^p dx$ *and* $x \in \mathfrak{D} \sim E$, $0 \leq n < q^N$ *implies* $|S_{n+1}f(x)| \leq C\,L\,y$.

If we establish (3.1) then since the estimate of $|E|$ does not depend on N, we see that (3.1) implies (1.3) with $C_p = C^2L(p)$ and in view of (3.2), (1.3) follows.

B. Development of assertions

In this section we develop assertions that will be used in §C to prove the Basic Lemma. As these assertions are stated (without proof) they are "starred". In §D we return to the assertions to prove them, "removing the stars".

There is a set $S^* \subset \mathfrak{P}^{-1}$ such that

$$(3.3^*) \qquad |S^*| \leq q\, y^{-p} \int_{\mathfrak{D}} |f(x)|^p dx$$

$$(3.4^*) \qquad \omega^* \not\subset S^* \Longrightarrow C_n(\omega^*;f) < y, \ \underline{\text{for all}}\ n = 0,1,\ldots\ .$$

For each positive integer k we define G_k^*, a collection of pairs $(n|\omega|,\omega^*)$ such that

$$(3.5^*) \qquad (n|\omega|,\omega^*) \in G_k^* \Longrightarrow n|\omega| \ \underline{\text{is a non-negative integer;}}$$
$$\omega^* \not\subset S^*;\ \omega^* \subset \mathfrak{P}^{-1};\ |\omega^*| > q^{-N+1}\ ;$$
$$C_{n[\omega]}(\omega^*;f) < q^{-k+1} y\ .$$

For each pair $(n|\omega|,\omega^*) \in G_k^*$ we define a partition of ω^*, $\Omega = \Omega(n|\omega|,\omega^*,k)$, into a finite union of mutually disjoint spheres $\tilde{\omega}^*$.

<u>Definition.</u> $\bar{\omega}^* \in \Omega = \Omega(n|\omega|,\omega^*,k)$ if $\bar{\omega}^* \subset \omega^*$ and,

(3.6) $C_{n[\tilde{\omega}]}(\tilde{\omega}^*;f) < q^{-k+1}y$ <u>for all</u> $\tilde{\omega}^*$ <u>such that</u>

$\bar{\omega}^* \subsetneq \tilde{\omega}^* \subset \omega^*$, <u>and</u>

$|\bar{\omega}^*| = q^{-N+1}$

(3.7) or

$|\bar{\omega}^*| > q^{-N+1}$ <u>and</u> $C_{n[\omega]}(\bar{\omega}^*;f) \geq q^{-k+1}y$.

This partition is obviously well defined as a proper partition of ω^*, That is $\bar{\omega}^* \in \Omega$ implies that $\bar{\omega}^* \subsetneq \omega^*$.

Given $\Omega = \Omega(n|\omega|,\omega^*,k)$ for $(n|\omega|,\omega^*) \in G_k^*$ we define $\mu^* = \mu^*(n|\omega|,\omega^*,k) \subset \omega^*$ such that

(3.8*) $|\mu^*| \leq q^{-5kL}|\omega^*|$, <u>and</u>

(3.9*) <u>If</u> $x \notin \mu^*$, $x \in \tilde{\tilde{\omega}}^* \subset \omega^*$, <u>and</u> $\omega^* \approx \tilde{\tilde{\omega}}^*$ <u>is a union of spheres in</u> Ω, <u>then there is a constant</u> $C_o > 0$, <u>independent of</u> y, p, N, f <u>and</u> k <u>such that</u>

$$|S_n^{\#}f(x;\omega^*) - S_n^{\#}f(x;\tilde{\tilde{\omega}}^*)| < C_o \, L \, k \, q^{-k+1}y \ .$$

To use (3.9) we need to avoid points x in the set

$$U^* = \cup_{k=1}^{\infty} \cup_{(n|\omega|,\omega^*)\in G_k^*} \mu^*(n|\omega|,\omega^*,k).$$

To estimate the measure of U^* we will show that

$$(3.10^*) \qquad \sum_{(n|\omega|,\omega^*)\in G_k^*} |\omega^*| \le q^{5kL-k} y^{-p} \int_{\Omega} |f(x)|^p dx .$$

Combining (3.8) and (3.10) we obtain:

$$(3.11) \qquad |U^*| \le \sum_{k=1}^{\infty} q^{-5kL} \sum_{(n|\omega|,\omega^*)\in G_k^*} |\omega^*|$$

$$\le \Big(\sum_{k=1}^{\infty} q^{-k}\Big) y^{-p} \int_{\Omega} |f(x)|^p dx \le y^{-p} \int_{\Omega} |f(x)|^p dx .$$

(3.12*) Let $n[\omega], \omega^*$ and k be given such that: $\omega^* \not\subset S^*$, $\omega^* \subset \mathfrak{P}^{-1}$, $|\omega^*| > q^{-N+1}$ and $C_{n[\omega]}(\omega^*;f) \ge q^{-k}y$, then there exists $(\bar{n},\bar{\omega}^*,\bar{k})$ such that: $\bar{n}[\omega] = n[\omega]$, $\bar{\omega}^* \supset \omega^*$, $1 \le \bar{k} \le k$, $(\bar{n}[\bar{\omega}],\bar{\omega}^*) \in G_{\bar{k}}^*$ and $C_{\bar{n}[\tilde{\omega}]}(\tilde{\bar{\omega}}^*;f) < q^{-\bar{k}+1}y$ for all $\tilde{\bar{\omega}}^*$ such that $\omega^* \subset \tilde{\bar{\omega}}^* \subset \mathfrak{P}^{-1}$.

In the special case $\omega^* = \mathfrak{P}^{-1}$ we have,

(3.13*) If $S^* \not\supset \mathfrak{P}^{-1}$ and $q^{-k}y \le C_n(\mathfrak{P}^{-1};f) < q^{-k+1}y$, then $(n,\mathfrak{P}^{-1}) \in G_k^*$.

We now use (3.12) to obtain:

(3.14) Suppose $n[\omega]$, ω^* and k are as in (3.12) and $(\bar{n},\bar{\omega}^*,\bar{k})$ is the triple given by that result. Let $\bar{\Omega} = \Omega(\bar{n}[\bar{\omega}],\bar{\omega}^*,\bar{k})$, x any

<u>fixed point in</u> ω^*, <u>and</u> $\omega^*(x)$ <u>be the sphere in</u> $\bar{\Omega}$ <u>that contains</u> x.

<u>Then</u>:

(a) $\omega^*(x) \subsetneqq \omega^*$

(b) $0 \le n < q^N \Longrightarrow \quad 0 \le \bar{n} < q^N$

(c) $x \in \omega^* \approx \mathcal{U}^*(\bar{n}|\bar{\omega}|,\bar{\omega}^*,\bar{k})$ <u>implies that</u>

$$|S_n^{\#} f(x;\omega^*)| \le |S_{\bar{n}}^{\#} f(x;\omega^*(x))| + q^{-\bar{k}+2} y + 2\,C_o\, L\, \bar{k}\, q^{-\bar{k}+1} y\ .$$

<u>Proof</u>. Since $\omega^*(x) \cap \omega^* \neq \phi$ then either $\omega^*(x) \subsetneqq \omega^*$ or $\omega^* \subset \omega^*(x)$. If the latter holds (3.7) implies that $C_{\bar{n}[\omega(x)]}(\omega^*(x);f) \ge q^{-\bar{k}+1} y$ (note that $|\omega^*(x)| \ge |\omega^*| > q^{-N+1}$). But the last condition of (3.12) says that $C_{\bar{n}[\omega(x)]}(\omega^*(x);f) < q^{-\bar{k}+1}$, which is a contradiction so $\omega^*(x)$ is properly contained in ω^*. This establishes (a).

If $0 \le n < q^N$, notice that $\bar{n}[\omega] = n[\omega]$ and $|\omega| \ge q^{-N+1}$. Thus, $|\omega|^{-1} = q^{N-\ell}$, $\ell \ge 1$; $n[\omega] \le q^{\ell}-1$, so $\bar{n} < \bar{n}[\omega]|\omega|^{-1} + |\omega|^{-1} < (q^{\ell}-1)q^{N-\ell} + q^{N-\ell} = q^N$. This establishes (b).

The estimate for $S_n^{\#} f(x,\omega^*)$ is obtained in two parts.

<u>First</u>: $$|S_n^{\#} f(x;\omega^*)| \le |S_{\bar{n}}^{\#} f(x;\omega^*)| + q\, C_{n[\omega]}(\omega^*;f) \qquad (2.8)$$

$$\le |S_{\bar{n}}^{\#} f(x;\omega^*)| + q^{-\bar{k}+2} y\ . \qquad (3.12)$$

<u>Second</u>: Let Ω be a partition of ω^* into a finite number of mutually disjoint spheres; let $\vec{\omega}^*$ be in the partition and let $\tilde{\omega}^*$

be a sphere such that $\overline{\omega}^* \subset \tilde{\omega}^* \subset \omega^*$. Then $\omega^* \approx \tilde{\omega}^*$ is a union of elements of Ω. Apply this observation to $\omega^*(x) \subset \omega^* \subset \overline{\omega}^*$ and we see that $\overline{\omega}^* \approx \omega^*$ is a union of spheres in $\overline{\Omega}(\overline{n}|\overline{\omega}|, \overline{\omega}^*, \overline{k})$ and it follows from (3.9) that if $x \notin \mathcal{U}^*(\overline{n}|\overline{\omega}|, \overline{\omega}^*, \overline{k})$, $x \in \omega^*(x) \subset \omega^* \subset \overline{\omega}^*$, then

$$|S^{\#}_{\overline{n}} f(x;\omega^*) - S^{\#}_{\overline{n}} f(x;\omega^*(x))|$$

$$\leq |S^{\#}_{\overline{n}} f(x;\overline{\omega}^*) - S^{\#}_{\overline{n}} f(x;\omega^*)| + |S^{\#}_{\overline{n}} f(x;\overline{\omega}^*) - S^{\#}_{\overline{n}} f(x;\omega^*(x))|$$

$$\leq 2\, C_o\, L\, \overline{k}\, q^{-\overline{k}+1} y .$$

We combine the two parts and the proof of (3.14) is complete.

C. Proof of the Basic Lemma

We now assemble a proof of (3.1) given (3.3) - (3.14).

The exceptional set is $E = S^* \cup U^*$. From (3.3) and (3.11) we have $|E| \leq C^p y^{-p} \int_{\Omega} |f(x)|^p dx$, where $C = \sup_{1<p<\infty} (1+q)^{1/p} = 1 + q$.

Fix $x \in \Omega \approx E$ and n such that $0 \leq n < q^N$ and consider $S_{n+1} f(x)$. We may assume that $c_n(\Omega;f) \neq 0$.

From (3.4) we have $C_n(\mathfrak{P}^{-1};f) < y$, otherwise $\mathfrak{P}^{-1} = E$ and there is nothing more to do. Thus, there is an integer $k_o \geq 1$ such that,

$$q^{-k_o} y \leq C_n(\mathfrak{P}^{-1};f) = |c_n(\Omega;f)| < q^{-k_o+1} y .$$

Using (3.13) we have $(n,\mathfrak{P}^{-1}) \in G^*_{k_o}$ so $\Omega_o = \Omega_o(n,\mathfrak{P}^{-1},k_o)$ is defined.

Let ω_1^* denote the sphere in Ω_o that contains x. From (3.9) we have:

$$(0)\quad \begin{cases} |S_{n+1}f(x)| \le |c_n(\mathfrak{D};f)| + |S_n^{\#}f(x;\mathfrak{P}^{-1})| \\ \qquad \le q^{-k_o+1}y + C_o\, L\, k_o q^{-k_o+1}y + |S_n^{\#}f(x;\omega_1^*)| \end{cases} .$$

Note that $\omega_1^* \subsetneqq \mathfrak{P}^{-1}$. If $|\omega_1^*| = q^{-N+1}$ we stop. If $|\omega_1^*| > q^{-N+1}$ we continue with a typical step:

From (3.7) $|\omega_1^*| > q^{-N+1} \Longrightarrow C_{n[\omega_1]}(\omega_1^*;f) \ge q^{-k_1}y$ for some k_1, $1 \le k_1 < k_o$. Apply (3.12) to obtain $(\bar{n}_1,\bar{\omega}_1^*,\bar{k}_1)$. Let ω_2^* be the sphere in $\Omega(\bar{n}_1|\bar{\omega}_1|,\omega_1^*,\bar{k}_1)$ which contains x. Then by (3.14)

$$(1)\quad \begin{cases} |S_n^{\#}f(x;\omega_1^*)| \le q^{-\bar{k}_1+2}y + 2\,C_o\,L\,\bar{k}_1 q^{-\bar{k}_1+1}y + |S_{\bar{n}_1}^{\#}f(x;\omega_2^*)| \\ \omega_2^* \subsetneqq \omega_1^*\,,\ 0 \le \bar{n}_1 < q^N\,,\ 1 \le \bar{k}_1 < k_o\,. \end{cases}$$

If $|\omega_2^*| = q^{-N+1}$ we stop. Otherwise we repeat the step (J-1)-times, $J \le N$ to obtain

$$(j)\quad \begin{cases} |S_{\bar{n}_{j-1}}^{\#}f(x;\omega_j^*)| \le q^{-\bar{k}_j+2}y + 2\,C_o\bar{k}_j q^{-\bar{k}_j+1}y + |S_{\bar{n}_j}^{\#}f(x;\omega_{j+1}^*)| \\ \omega_{j+1}^* \subsetneqq \omega_j^*\,,\ 0 \le n_j < q^N\,,\ 1 \le \bar{k}_j < \bar{k}_{j-1}\,, \end{cases}$$

$j = 2,3,\ldots,J, |\omega_{J+1}| = q^{-N+1}, x \in \omega^*_{J+1}$.

Then $\bar{n}_J[\omega_{J+1}] = 0$ so by (2.8), (3.4) and the fact that $S^{\#}_o f \equiv 0$ gives us,

$$(J+1) \qquad |S^{\#}_{\bar{n}_J} f(x;\omega^*_{J+1})| \leq q\, C_o(\omega^*_{J+1};f) < q\, y .$$

Combining (0),(1),...,(J),(J+1) we get

$$|S_{n+1}f(x)| \leq \left\{\left(\sum_{k=1}^{\infty} q^{-k+2}\right) + 2\, C_o\, L\left(\sum_{k=1}^{\infty} k\, q^{-k+1}\right) + q\right\} y .$$

This completes the proof of (3.1) modulo the establishing of the assertions: (3.3), (3.4), (3.5), (3.8), (3.9), (3.10), (3.12) and (3.13).

D. Proof of assertions

Proof of (3.3) and (3.4) (The set S^*). Let S be the union of all spheres $\omega \subset \mathfrak{D}$ such that

$$(3.15) \qquad \int_{\omega} |f(x)|^p dx \geq y^p |\omega| .$$

It is easy to see that S can be written a countable union of mutually disjoint spheres that satisfy (3.15).

Hence,

(3.16) $$|S| \le y^{-p} \int_{\mathcal{D}} |f(x)|^p dx \text{ , and}$$

(3.17) $$\omega \not\subset S \Longrightarrow |c_n(\omega;f)| \le \frac{1}{|\omega|} \int_{\omega} f(x)dx$$

$$\le \left[\frac{1}{|\omega|} \int_{\omega} |f(x)|^p dx\right]^{1/p} < y \qquad (3.15).$$

Let $S^* = \bigcup_{\omega \subset S} \omega^*$. From (3.16) we have

(3.3) $$|S^*| \le q|S| \le q\, y^{-p} \int_{\mathcal{D}} |f(x)|^p dx \text{ .}$$

If $\omega^* \not\subset S^*$ then $\bar{\omega} \subset S$ for all $\bar{\omega} \ni (\bar{\omega})^* = \omega^*$. Thus,

(3.4) $$\omega^* \not\subset S^* \Longrightarrow c_n(\omega^*;f) < y \quad \text{all } n \quad (3.17).$$

<u>Proof of (3.5), (3.12) and (3.13) (The pairs G_k^* and changing pairs).</u>

We construct polynomials $p_k(x;\omega)$, for k a positive integer; ω a shere, $\omega \subset \mathcal{B}^{-1}$. The definition is inductive. Define: $p_k(x;\mathcal{B}^{-1}) \equiv 0$.

For k fixed, $\omega \subset \mathcal{D}$ we assume that $p_k(x;\omega^*)$ is defined, $|\omega| \ge q^{-N+1}$. Recall that ω is reached by a unique chain

$\{\omega_j\}_{j=0}^{J}$ with $\omega_o = \mathfrak{B}^{-1}$, $\omega_J = \omega$ and $\omega_{j-1} = \omega_j^*$, $1 \leq j \leq J$.

Let $G_k(\omega) = \{(n,\omega): |c_n(\omega;f-p_k(\cdot;\omega^*))| \geq q^{-k}y\}$.

$$p_k(x;\omega) = p_k(x;\omega^*) + \sum_{(n,\omega)\in G_k(\omega)} c_n(\omega;f-p_k(\cdot;\omega^*))\chi_{n|\omega|^{-1}}(x) .$$

For $\omega \subset \mathfrak{D}$

(3.18) $\qquad (n,\omega) \in G_k(\omega) \Rightarrow |c_n(\omega;f-p_k(\cdot;\omega^*))| \geq q^{-k}y$

(3.19) $\qquad \forall(m,\omega),\ |c_m(\omega;f-p_k(\cdot;\omega))| < q^{-k}y$.

We may write,

$$p_k(x;\omega) = \sum_{\substack{\omega' \\ \omega\subset\omega'\subset\mathfrak{D}}} \sum_{(n'|\omega'|,\omega')\in G_k(\omega')} c_{n'|\omega'|}(\omega';f-p_k(\cdot;(\omega')^*))\chi_{n'}(x) .$$

From this representation we see:

(3.20) $\quad$ <u>If</u> $p_k(x;\omega)$ <u>contains a term</u> $c\,\chi_{n'}$, <u>then there exists</u> ω' <u>such that</u> $\omega \subset \omega' \subset \mathfrak{D}$ <u>and</u> $(n'|\omega'|,\omega') \in G_k(\omega')$.

Let G_k^* be the collection of pairs $(n|\omega|,\omega^*)$ such that $(n|\omega|,\omega) \in G_k(\omega)$, $|\omega| \geq q^{-N+1}$, $\omega^* \subset S^*$ and $C_{n|\omega|}(\omega^*;f) < q^{-k+1}y$.

(3.5) $\quad$ The conditions of (3.5*) are now met.

(3.13) $\quad$ The conditions of (3.13*) are also met.

(3.21) <u>If</u> $|c_m(\omega;f)| \geq q^{-k}y$ <u>then there are</u> n', ω' <u>such that</u> $\omega \subset \omega' \subset \mathfrak{D}$, $n'[\omega] = m$ <u>and</u> $(n'|\omega'|, \omega') \in G_k(\omega')$.

<u>Proof of (3.21)</u>. The result will follow from (3.20) if $p_k(x;\omega)$ contains a term $c\,x_{n'}$ with $n'[\omega] = m$. If $p_k(x;\omega)$ contains no such term then $c_m(\omega;p_k(\cdot;\omega)) = 0$, and so $|c_m(\omega;f - p_k(\cdot,\omega))|$ $= |c_m(\omega;f)| \geq q^{-k}y$, which contradicts (3.19).

(3.12) <u>If</u> $n[\omega], \omega^*$ and k <u>are given so that</u>: $\omega^* \not\subset S^*$, $\omega^* \subset \mathfrak{P}^{-1}$; $|\omega^*| > q^{-N+1}$ <u>and</u> $C_{n[\omega]}(\omega^*;f) \geq q^{-k}y$; <u>then there exists</u> $(\bar{n}, \bar{\omega}^*, \bar{k})$ <u>such that</u>: $\bar{n}[\omega] = n[\omega]$, $\bar{\omega}^* \supset \omega^*$, $1 \leq \bar{k} \leq k$, $(\bar{n}[\bar{\omega}], \bar{\omega}^*) \in G^*_{\bar{k}}$ <u>and</u> $C_{\bar{n}[\tilde{\omega}]}(\tilde{\tilde{\omega}}^*;f) < q^{-\bar{k}+1}$ <u>for all</u> $\tilde{\tilde{\omega}}^*$ <u>such that</u> $\omega^* \subset \tilde{\tilde{\omega}}^* \subset \mathfrak{P}^{-1}$.

<u>Proof of (3.12)</u>. Let Σ denote the collection of triples (n', ω', k') such that:

(i) $1 \leq k' \leq k$; (ii) $\omega' \subset (\omega')^* \subset \mathfrak{P}^{-1}$; (iii) $n'[\omega] = n[\omega]$;

(iv) $(n'|\omega'|, \omega') \in G_k(\omega')$.

Let us check that Σ is not empty: Since $C_{n[\omega]}(\omega^*;f) \geq q^{-k}y$, there is an ω, $(\omega)^* = \omega^*$ such that $|c_{n[\omega]}(\omega;f)| \geq q^{-k}y$. By (3.21) there are n', ω' such that $\omega \subset \omega' \subset \mathfrak{D}$, $n'[\omega] = n[\omega]$ and $(n'|\omega'|, \omega') \in G_k(\omega')$, so $(n', \omega', k) \in \Sigma$ and Σ is not empty.

Let $(\bar{n}, \bar{\omega}, \bar{k})$ be an element of Σ with $\bar{k}$ minimal. We claim that $(\bar{n}, \bar{\omega}^*, \bar{k})$ is the required triple. We only need to verify the last condition of (3.12), the other parts being obvious.

Thus, suppose $\omega^* \subset \tilde{\omega}^* \subset \mathfrak{B}^{-1}$, we will show that $C_{\bar{n}[\tilde{\omega}]}(\tilde{\omega}^*;f) \geq q^{-\bar{k}+1}y$ leads to a contradiction. If that inequality holds, since $\tilde{\omega}^* \notin S^*$, we know that $C_{\bar{n}[\tilde{\omega}]}(\tilde{\omega}^*;f) < y$ and so $(\bar{k}-1) \geq 1$. Apply (3.21) with $\bar{n}[\tilde{\omega}],\tilde{\omega}$, and $(\bar{k}-1)$ and obtain $(n'|\omega'|,\omega') \in G_{\bar{k}-1}(\omega')$.

We have, $1 \leq \bar{k}-1 < \bar{k} \leq k$. $\tilde{\omega} \subset \omega'$ and $\omega^* \subset \tilde{\omega}^*$ implies $\omega^* \subset (\omega')^* \subset \mathfrak{B}^{-1}$. Next, $n'[\tilde{\omega}] = \bar{n}[\tilde{\omega}]$, $|\tilde{\omega}| \geq |\omega|$ implies $n'[\omega] = \bar{n}[\omega]$. Since $\bar{n}[\omega] = n[\omega]$ we have $n'[\omega] = n[\omega]$. Since $(n'|\omega'|\omega') \in G_{\bar{k}-1}(\omega')$ we have $(n',\omega',\bar{k}-1) \in \Sigma$, a contradiction to the minimality of $\bar{k}$. The proof of (3.12) is complete.

<u>Proof of (3.10) (How much $|\omega^*|$ is there in G_k^*).</u>

We are going to estimate $\sum_{(n,\omega^*)\in G_k^*} |\omega^*|$. We will obtain an estimate of $\sum_{j=0}^{N-1} \sum_{\substack{(n,\omega)\in G_k(\omega) \\ |\omega|=q^{-j}}} |\omega|$. When we multiply this estimate by q, we will have the required estimate.

Let $a_n(\omega) = c_n(\omega;f-p_k(\cdot;\omega^*))$. Using Plancherel's formula, (2.5), we obtain:

$$\sum_{|\omega|=q^{-N+1}} \int_{\omega} |f(x)-p_k(x;\omega)|^2 dx$$

$$= \sum_{|\omega|=q^{-N+1}} \int_{\omega} \Big| (f(x)-p_k(x;\omega^*)) - \sum_{(n,\omega)\in G_k(\omega)} a_n(\omega) \chi_{nq^{N-1}}(x) \Big|^2 dx$$

$$= \sum_{|\omega|=q^{-N+1}} \int_{\omega} |f(x)-p_k(x;\omega^*)|^2 dx - \sum_{\substack{(n,\omega)\in G_k(\omega) \\ |\omega|=q^{-N+1}}} |a_n(\omega)|^2 |\omega|$$

$$= \sum_{|\omega|=q^{-N+2}} \int_{\omega} |f(x)-p_k(x;\omega)|^2 dx - \sum_{\substack{(n,\omega)\in G_k(\omega) \\ |\omega|=q^{-N+1}}} |a_n(\omega)|^2 |\omega| .$$

We repeat this calculation (N-1)-times and obtain,

$$0 \leq \sum_{|\omega|=q^{-N+1}} \int_{\omega} |f(x)-p_k(x;\omega)|^2 dx$$

$$= \int_{\mathfrak{D}} |f(x)|^2 dx - \sum_{j=0}^{N-1} \sum_{\substack{(n,\omega)\in G_k(\omega) \\ |\omega|=q^{-j}}} |a_n(\omega)|^2 |\omega| .$$

Thus,

$$\sum_{j=0}^{N-1} \sum_{\substack{(n,\omega)\in G_k(\omega) \\ |\omega|=q^{-j}}} |a_n(\omega)|^2 |\omega| \leq \int_{\mathfrak{D}} |f(x)|^2 dx .$$

From (3.18) we have $|a_n(\omega)| \geq q^{-k}y$. Hence

$$\sum_{j=0}^{N-1} \sum_{\substack{(n,\omega)\in G_k(\omega) \\ |\omega|=q^{-j}}} |\omega| \leq q^{2k} y^{-2} \int_{\mathcal{D}} |f(x)|^2 dx ,$$

and so

$$(3.22) \qquad \sum_{(n|\omega|,\omega^*)\in G_k^*} |\omega^*| \leq q^{2k+1} y^{-2} \int_{\mathcal{D}} |f(x)|^2 dx .$$

Since $L(2) = 8$ and $2k+1 < 5\cdot 8\cdot k-k = 39k$, $k \geq 1$, we see that for $p = 2$ we are done with the proof of (3.10).

For $p \neq 2$ we need an additional observation:

(3.23) <u>If</u> $G_k(\omega)$ <u>contains a pair</u> (n,ω) <u>with</u> $\omega \notin S$ <u>then</u> $y^{-2} \leq q^{kL} y^{-p}$, <u>where</u> $L = L(p) = [2p^2/(p-1)] + 1$.

Before presenting a proof we show how it is applied. If G_k^* is empty we are done with (3.10). If G_k^* is not empty then there is a $G_k(\omega)$ such that $\omega \notin S$ and $(n,\omega) \in G_k(\omega)$ and so $y^{-2} \leq q^{kL} y^{-p}$. Now use (3.22) and recall that f is the characteristic function of a set. Thus,

$$\sum_{(n,\omega^*)\in G_k^*} |\omega^*| \leq q^{2k+1} y^{-2} \int_{\mathcal{D}} |f(x)|^2 dx$$

$$\leq q^{2k+1} q^{kL} y^{-p} \int_{\mathcal{D}} |f(x)|^p \, dx$$

$$= q^{(2+L)k+1} y^{-p} \int_{\mathcal{D}} |f(x)|^p dx .$$

Since, $(2+L)k+1 \leq (5L-1)k$, $k \geq 1$,

$$(3.10) \qquad \sum_{(n,\omega^*)\in G_k^*} |\omega^*| \leq q^{5Lk-k} y^{-p} \int_{\mathcal{D}} |f(x)|^p dx .$$

This proves (3.10).

Proof of (3.23). If $(n,\omega) \in G_k(\omega)$ then there is a pair (m',ω') such that $\omega' \supset \omega$ and $|c_m(\omega';f)| \geq q^{-k}y$. For suppose $|c_m(\omega';f)| < q^{-k}y$ for all (m,ω') such that $\omega' \supsetneq \omega$. Then $p_k(x;\omega^*) \equiv 0$, so $|c_n(\omega;f)| = |c_n(\omega;f-p_k(\cdot;\omega^*))| \geq q^{-k}y$ and (n,ω) is the required pair. Fix such a pair (m,ω'). Since $\omega \notin S$, $\omega' \notin S$.

Case I. $1 < p < 2$. f is the characteristic function of a set. Using (3.15) and $\omega' \notin S$ we have,

$$q^{-k}y \leq |c_m(\omega';f)| \leq \frac{1}{|\omega'|} \int_{\omega'} |f(x)| dx \leq \frac{1}{|\omega'|} \int_{\omega'} |f(x)|^p dx \leq y^p .$$

Thus, $y^{p-1} \geq q^{-k}$. We want to show that $y^{2-p} \geq q^{-kL}$.

But, $y^{2-p} \geq q^{-\frac{2-p}{p-1}\cdot k}$. But $(2-p)/(p-1) \leq 2p^2/(p-1) \leq L$ for $1 < p < 2$, so $y^{2-p} \geq q^{-kL}$.

Case II. $2 < p < \infty$. $q^{-k}y \leq |c_n(\omega';f)| \leq \frac{1}{|\omega'|}\int_{\omega}|f(x)|dx \leq 1$, since f is the characteristic function of a set. Thus, $y \leq q^k$, so $y^{p-2} \leq q^{k(p-2)}$. But $(p-2) \leq L$, if $p > 2$, so $y^{p-2} \leq q^{kL}$.

The proof of (3.23) is complete.

Proof of (3.8) and (3.9) (The set μ^*)

For $(n|\omega|,\omega^*) \in G_k^*$ we construct the partition $\Omega = \Omega(n|\omega|,\omega^*,k)$ of ω^*. We define g on ω^* by

$$g(t) = \frac{1}{|\overline{\omega}|}\int_{\overline{\omega}} f(z)\overline{\chi}_n(z)dz, \quad t \in \overline{\omega},\ \overline{\omega} \in \Omega .$$

$$g(t) = |c_{n[\overline{\omega}]}(\overline{\omega};f)| \leq C_{n[\overline{\omega}]}(\overline{\omega}^*;f).$$

Since $\overline{\omega} \in \Omega$, $\overline{\omega} \subsetneq \overline{\omega}^* \subset \omega^*$, (3.6) implies

$$\|g\|_{\infty,\omega^*} < q^{-k+1}y \tag{3.25}$$

(3.26) Suppose $x \in \omega^*$, $\tilde{\tilde{\omega}} \subset \omega^*$ is such that $\omega^* \approx \tilde{\tilde{\omega}}^*$ is a union of spheres in Ω and $x \in \tilde{\tilde{\omega}}^*$. Then

$$|S_n^{\#}f(x;\omega^*) - S_n^{\#}(x;\tilde{\tilde{\omega}}^*)| \leq T_n^{\#}g(x;\omega^*) .$$

<u>Proof</u>.

$$|S_n^{\#}f(x;\omega^*) - S_n^{\#}f(x;\tilde{\tilde{\omega}}^*)|$$

$$= |\int_{\omega^* \approx \tilde{\tilde{\omega}}^*} f(t)\bar{\chi}_n(t)D_n^{\#}(x-t)dt|$$

$$= |\int_{\omega^* \sim \tilde{\tilde{\omega}}^*} g(t)D_n^{\#}(x-t)dt$$

$$+ \int_{\omega^* \approx \tilde{\omega}^*} (f(t)\bar{\chi}_n(t)-g(t))D_n^{\#}(x-t)dt|$$

$\omega^* \approx \tilde{\tilde{\omega}}^*$ is the union of mutually disjoint spheres $\{\omega'\}$, such that $x \in \omega'$, $\omega' \in \Omega$. From (2.7) we see that $D_n^{\#}(x-t)$ is constant on ω', and from the definition of g we see that $\int_{\omega'}(f(t)\bar{\chi}_n(t)-g(t))dt = 0$, so the second term in the sum is zero, and the result follows.

Now let $A > 0$ be the constant of (2.9) (which depends only on K) and let $C_o > 0$ be such that $A\, C_o \geq (2 + 5 \log q)$. For $(n|\omega|,\omega^*) \in G_k^*$ let $\mu^* = \mu^*(n|\omega|,\omega^*,k) =$

$$= \{x \in \omega^*: T_n^{\#}\, g(x;\omega^*) > C_o L k q^{-k+1} y\}.$$

From (2.9) and (3.25)

$$|\mu^*| \leq e^2 \exp\{-A\, C_o\, L\, k\, q^{-k+1} y/\|g\|_{\infty,\omega^*}\}|\omega^*|$$

$$\leq e^2 \exp\{-A\, C_o\, L\, k\}|\omega^*|$$

$$\leq e^2 \exp\{-2kL\} \exp\{-5k\, L \log q\}\, |\omega^*| \tag{3.8}$$

$$\leq q^{-5kL}|\omega^*| \;.$$

(3.9) If we combine (3.26) with the definition of μ^* we see that (3.9) is established.

4. Notes for Chapter VIII

The general reference for this chapter is Hunt and Taibleson [1].

Carleson [1] established the conjecture of Lusin; namely, that if $f \in L^2(0,2\pi)$, its trigonometric Fourier series converged a.e. Shortly thereafter Billard [1] extended this result to Walsh-Fourier series, an Hunt [2] extended it to $f \in L^p(0,2\pi)$, $p > 1$, for trigonometric Fourier series. Sjölin [1] made the extension of the L^p result to Walsh-Fourier series and then Hunt and Taibleson extended the results to thos of this section which includes the result of Sjölin as a special case.

The driving idea of Carleson's proof was to show that $f \to \mathfrak{M}f$ (see §1 for the definition) was of weak type (2,2). That is, if $f \in L$ then for each $y > 0$, $|\{x : \mathfrak{M}f(x) > y\}| \leq Cy^{-2}\|f\|_2^2$.

No proof has been given to show, directly, that the map is of weak type (p,p), $p \neq 2$, but Hunt showed that it was of restricted weak type (p,p), $1 < p < \infty$. That is, if f is the characteristic function of a measurable set F, then $|\{x : \mathfrak{M}f(x) > y\}| \leq Cy^{-p}|F|$.

When a sublinear map is of restricted weak type for a range of values, one can interpolate and obtain the strong type result for intermediate values and the conclusion of Theorem (1.1) results. The notion of restricted weak type was introduced by Stein and Weiss [1] and their interpolation theorem will suffice for our conclusion. A more general treatment of such interpolation theorems in the setting of Lorentz spaces, L(p,q), can be found in Hunt [1].

Bibliography

N. Aronszajn and K. T. Smith

[1] "Theory of Bessel potentials I," Ann. Inst. Fourier 11(1961), 385-475.

A. Benedek and R. Panzone

[1] "The spaces L^p with mixed norm," Duke Math. J. 28(1961), 301-324.

P. Billard

[1] "Sur la convergence presque partout des serie de Fourier-Walsh de fonctions de l'espace $L^2(0,1)$," Studia Math. 28(1966/67), 363-388.

N. Bourbaki

[1] Algebre Commutative, Paris, 1964.

D. L. Burkholder

[1] "Maximal inequalities as necessary conditions for a.e. convergence," Z. Wahrschein. Verw. Gebeite 3(1964), 75-88.

[2] "Martigale transforms," Ann. Math. Statist. 37(1966), 1494-1504.

A. P. Calderón

[1] "On the behaviour of harmonic functions on the boundary," Trans. Amer. Math. Soc. 68(1950), 47-54.

[2] "On a theorem of Marcinkiewicz and Zygmund," Trans. Amer. Math. Soc. 68(1950), 55-61.

[3] "Lebesgue spaces of differentiable functions and distributions," Proc. Symp. in Pure Math. 5(1961), 33-49.

A. P. Calderón and A. Zygmund

[1] "On the existence of certain singular integrals," Acta Math. 88(1952), 85-139.

L. Carleson

[1] "On the convergence and growth of partial sums of Fourier series," Acta Math. 116(1966), 135-157.

J-A. Chao

[1] "H^p-spaces of conjugate systems on local fields," Studia Math. 49(1974), 267-287.

[2] "Maximal singular integral transforms on local fields," to appear in Proc. Amer. Math. Soc.

J-A. Chao and M. H. Taibleson

[1] "A sub-regularity inequality for conjugate systems on local fields," Studia Math. 46(1973), 249-257.

R. R. Coifman and G. Weiss

[1] "On subharmonicity inequalities involving solutions of generalized Cauchy-Riemann equations," Studia Math. 36(1970), 77-83.

J. L. Doob

[1] Stochastic Processes, New York, 1953.

R. E. Edwards

[1] Fourier Series, Vols. I and II, New York, 1967.

R. E. Edwards and E. Hewitt

[1] "Pointwise limits for a sequence of convolution operators," Acta Math. 113(1965), 181-218.

N. J. Fine

[1] "On the Walsh functions," Trans. Amer. Math. Soc. 65(1949), 372-414.

[2] "Cesàro summability of Walsh-Fourier series," Proc. Nat. Acad Sci. U.S.A. 41(1955), 588-591.

I. M. Gelfand and M. T. Graev

[1] "Representations of a group of the second order with elements from a locally compact field and special functions on locally compact fields," Uspehi Mat. Nauk, Russian Math. Surveys 18(1963), 29-100.

R. F. Gundy

[1] "Martingale theory and convergence of certain orthogonal series," Trans. Amer. Math. Soc. 124(1966), 228-248.

[2] "A decomposition for L^1 bounded martingales," Ann. Math. Statist. 39(1968), 134-138.

G. H. Hardy, J. E. Littlewood and G. Polyà

[1] Inequalities, Cambridge, 1959.

L. H. Harper

[1] "Capacities of sets and harmonic analysis on the group 2^{ω}," Trans. Amer. Math. Soc. 126(1967), 303-315.

E. Hewitt and K. Ross

[1] Abstract Harmonic Analysis I, Heidelberg, 1963.

L. Hörmander

[1] "Estimates for translation invariant operators in L_p spaces," Acta Math. 104(1960), 93-140.

R. A. Hunt

[1] "On L(p,q) spaces," Enseignment Math. 12(1966), 249-276.

[2] "On the convergence of Fourier series," Conf. on Orthogonal Expansions and their Continuous Analogues, Southern Ill. Univ. Press, Edwardsville, 1967.

R. A. Hunt and M. H. Taibleson

[1] "Almost everywhere convergence of Fourier series on the ring of integers in a local field," SIAM J. Math Analysis 2(1971), 607-625.

S. Lang

[1] Algebraic Numbers, Reading, Mass., 1964.

J. E. Littlewood and R. E. A. C. Paley

[1] "Theorems on Fourier series and power series," J. London Math. Soc. 6(1931), 230-233.

A. E. Monna

[1] "Linear topological vector spaces over non-archimedian valued fields," Proc. Conf. on Local Fields (Dreibergen) 56-65, Berlin, 1967.

R. O'Neil

[1] "Fractional integration on Orlicz spaces, I," Trans. Amer. Math. Soc. 115(1965), 300-328.

C. W. Onneweer

[1] "Absolute convergence of Fourier series on certain groups," Duke Math. J. 39(1972), 599-609.

R. E. A. C. Paley

[1] "A remarkable system of orthogonal functions," Proc. London Math. Soc. 34(1932), 241-279.

K. Phillips

[1] "Hilbert transforms for the p-adic and p-series fields," Pacific J. Math. 23(1967), 329-347.

K. Phillips and M. H. Taibleson

[1] "Singular integrals in several variables over a local field," Pacific J. Math. 30(1969), 209-231.

P. J. Sally and M. H. Taibleson

[1] "Special functions on locally compact fields," Acta Math. 116(1966), 279-309.

L. Schwartz

[1] Theorie des Distributions, Paris, 1951.

W. H. Schikof

[1] Non-Archimedian Harmonic Analysis (dissertation), Matematisch Institut, Katholieke Universiteit, Nijmegen, The Netherlands, 1967.

P. Sjölin

[1] "An inequality of Paley and convergence e. of Walsh-Fourier series," Ark. Mat. 7(1969), 551-570.

E. M. Stein

[1] "On the functions of Littlewood-Paley, Lusin and Marcinkiewicz," Trans. Amer. Math. Soc. 88 (1958), 430-466.

[2] Singular integrals and Differentiability Properties of Functions, Princeton, 1970.

[3] Topics in Harmonic Analysis Related to Littlewood-Paley Theory, Ann. of Math. Studies, No. 63, Princeton, 1970.

[4] "The characterization of functions arising as potentials," Bull. Amer. Math. Soc. 67(1961), 102-104.

E. M. Stein and G. Weiss

[1] "An extension of a theorem of Marcinkiewicz and some applications," J. Math. Mech. 8(1959), 263-284.

[2] Introduction to Fourier Analysis on Euclidean Space, Princeton, 1971.

[3] "Fractional integrals on n-dimensional Euclidean space," J. Math. Mech. 7(1958), 503-514.

E. Study

[1] "Ueber eine besondere Classe von Functionen einer reelen Verandlichen," Math. Ann. 47(1896), 298-316.

M. H. Taibleson

[1] "Fourier coefficients of functions of bounded variation," <u>Proc. Amer. Math. Soc</u>. 18(1967), 766.

[2] "Fourier series on the ring of integers in a p-series field," <u>Bull. Amer. Math. Soc</u>. 73(1967), 623-629.

[3] "Harmonic analysis on n-dimensional vector spaces over local fields, I. Basic results on fractional integration," <u>Math. Ann</u>. 176(1968), 191-207.

[4] "________, II. Generalized Gauss kernels and the Littlewood-Paley function," <u>Math. Ann</u>. 186(1970), 1-19.

[5] "________, III. Multipliers," <u>Math. Ann</u>. 187(1970), 259-271.

[6] "On the theory of Lipschitz spaces of distributions on Euclidean n-space, I. Principal properties," <u>J. Math. Mech</u>. 13(1964), 407-480.

N. Ja. Vilenkin

[1] "On a class of complete orthonormal systems," <u>Amer. Math. Soc. Transl</u>.(2) 28(1963), 1-35.

S. Wainger

[1] "Special trigonometrical series in k-dimensions," <u>Memoirs Amer. Math. Soc</u>. 59(1965), 1-102.

J. L. Walsh

[1] "A closed set of orthogonal functions," Amer. J. Math. 55(1923), 5-24.

D. Waterman

[1] "On convergence of Fourier series of functions of generalized bounded variation," Studia Math. 44(1972), 107-117.

A. Weil

[1] Basic Number Theory, New York, 1967.

A. Zygmund

[1] Trigonometrical Series, Cambridge, 1959.

[2] "On the convergence and summability of power series on the circle of convergence (I)," Fund. Math. 30(1928), 170-196.

Library of Congress Cataloging in Publication Data

Taibleson, M H 1929-
Fourier analysis on local fields.

(Mathematical notes ; no. 15)
Bibliography: p.
1. Fields, Algebraic. 2. Fourier analysis.
I. Title. II. Series: Mathematical notes
(Princeton, N. J.) ; no. 15.
QA247.T28 1975 512'.3 74-32047
ISBN 0-691-08165-4